Henry Sukardi

MEMS Computer Vision and Robotic Manipulation System

Henry Sukardi

MEMS Computer Vision and Robotic Manipulation System

LAP LAMBERT Academic Publishing

Impressum / Imprint
Bibliografische Information der Deutschen Nationalbibliothek: Die Deutsche
Nationalbibliothek verzeichnet diese Publikation in der Deutschen
Nationalbibliografie; detaillierte bibliografische Daten sind im Internet über
http://dnb.d-nb.de abrufbar.
Alle in diesem Buch genannten Marken und Produktnamen unterliegen
warenzeichen-, marken- oder patentrechtlichem Schutz bzw. sind
Warenzeichen oder eingetragene Warenzeichen der jeweiligen Inhaber. Die
Wiedergabe von Marken, Produktnamen, Gebrauchsnamen, Handelsnamen,
Warenbezeichnungen u.s.w. in diesem Werk berechtigt auch ohne besondere
Kennzeichnung nicht zu der Annahme, dass solche Namen im Sinne der
Warenzeichen- und Markenschutzgesetzgebung als frei zu betrachten wären
und daher von jedermann benutzt werden dürften.

Bibliographic information published by the Deutsche Nationalbibliothek: The
Deutsche Nationalbibliothek lists this publication in the Deutsche
Nationalbibliografie; detailed bibliographic data are available in the Internet
at http://dnb.d-nb.de.
Any brand names and product names mentioned in this book are subject to
trademark, brand or patent protection and are trademarks or registered
trademarks of their respective holders. The use of brand names, product
names, common names, trade names, product descriptions etc. even without
a particular marking in this work is in no way to be construed to mean that
such names may be regarded as unrestricted in respect of trademark and
brand protection legislation and could thus be used by anyone.

Coverbild / Cover image: www.ingimage.com

Verlag / Publisher:
LAP LAMBERT Academic Publishing
ist ein Imprint der / is a trademark of
OmniScriptum GmbH & Co. KG
Heinrich-Böcking-Str. 6-8, 66121 Saarbrücken, Deutschland / Germany
Email: info@lap-publishing.com

Herstellung: siehe letzte Seite /
Printed at: see last page
ISBN: 978-3-659-77828-5

Zugl. / Approved by: Victoria, University of Victoria, Diss., 2015

Abstract

Supervisory Committee
Nikolai Dechev, Mechanical Engineering Department
Supervisor
Afzal Suleman, Mechanical Engineering Department
Departmental Member

The main focus of this thesis is to explore the possibilities of micro-assembly by using computer vision as a main feedback tool for automation. In the current industry and research, most micro-assembly and micro-manipulation systems have utilized tele-operated approaches as a main methodology to achieving successful results. This is very labour intensive and not a cost effective process. Some have used computer vision to an extensive range to achieve successful manipulation for complex micro-parts.

Several improvements were made on the robotic system used in this work, to manipulate micro parts. Initially, the development of a servo-based machine utilizing encoders to attain high accuracy with the ability to move in higher resolutions, was done. Ultimately, work with a stepper motor based system was used given challenges with the previous system.

An optical microscopy system with high resolution, low image distortion and short depth of field, high frame rate and low processing latency image sensor, is used as a real time feedback tool for the vision acquisition and manipulation process. All of these have to be coupled by a computer, to process the motion and vision control simultaneously.

In addition, this work involved the design and development of MEMS components which were fabricated on a MEMS chip. These components were manipulated, in order to make automated assembly possible. A great deal of research and development was done to scrutinize previous chip designs and possible ways to improve the existing design. The chip components have incorporated changes in geometry and tolerances to improve the probability of success of gripping and joining. The MEMS chip also has a unique form of surface markers as reference geometry for the computer vision to calibrate its position in space and have a set origin. The chip has also been designed in a way that streamlines the efficiency of the micro-assembly process and minimizes the number of movements required in order to achieve successful assembly.

Several computer vision techniques were learned and explored to determine the best strategy to acquire and identify micro parts on the chip, and to manipulate them accordingly. These techniques lay the foundations of what future work will be based on.

Together, the computer vision techniques, the robotic system and the MEMS chips were integrated, leading to a completely automated gripping sequence that has not been done previously.

Table of Contents

LIST OF TABLES

LIST OF FIGURES

Nomenclature

Acronyms

MEMS	**Microelectromechanical Systems**
CAD	**Computer Aided Design**
DOF	**Degree Of Freedom**
CCD	**Charged Coupled Device**
CMOS	**Complimentary Metal Oxide Semiconductor**
ROI	**Region of Interest**
PCB	**Printed Circuit Board**
LoG	**Laplacian of Gaussian**
FPS	**Frames Per Second**
SEM	**Scanning Electron Microscope**
RGB	**Red Green Blue**

ACKNOWLEDGEMENTS

I would like to thank my parents for their support and wisdom throughout my program and all my siblings (Hendrian, Henderson and Hendrick) for their morale support throughout my program.

I would also like to thank Dr. Dechev for supporting this project and his valuable recommendations and advice for making this possible.

Chapter 1

INTRODUCTION

1.1 Statement of the Problem

MEMS (Micro-electromechanical Systems) devices have been conventionally fabricated using photolithography processes similar to CMOS (Complementary metal-oxide Semiconductor) technology. This top to bottom form of fabrication process has produced many of the world's remarkable MEMS devices. To improve the current fabrication process, it is suggested to have the ability to pick and place MEMS parts, and to join them together to make more useful devices. This increases the possibilities of the types of devices that can be derived using this type of fabrication process.

The current problem with MEMS micromanipulation, is the requirement to use human tele-operated systems to successfully grip and place micro-parts on a chip and join them together on a chip substrate. However, this is a very labour intensive process and an expensive one. In order to be able to make micro-assembly a viable industry standard, automation of the assembly process is required. The proposed means of automating micro-assembly would use computer vision as a main feedback tool, along with a highly precise servoing robotic system as a means to assemble the parts together.

1.2 The State-of-the-Art

MEMS micro-manipulation technology requires several key technologies to make it work. The technologies involved are MEMS chip design methodologies and fabrication, motion control, digital imaging, microscopy and computer vision. MEMS chips are fabricated in foundries using specific tech files, and in this particular instance, the PolyMUMPS process. For every specific

1

fabrication technology, designers have to work within the confines of the technology and the tolerances associated with the fabrication process. Accurate motion control has to be achieved in order to manipulate parts on the micron-nano level, this can be done either with an advanced and highly precise stepper motor or servo motors with encoder feedback. Image capturing of MEMS parts has to be done with a microscope with high optical resolution coupled with an image sensor (camera) with an equally high digital resolution to discern details on the chip. Computer vision techniques would have to be implemented to post process the image capture in order to recognize parts on the chip and provide feedback to the motion controller to successfully assemble the parts together.

At the present moment, there exist various micro-manipulation technologies from purely tele-operated systems to simple automated micromanipulators (end-effectors) that assemble micro spherical parts to fixed position in space. This section explores the various technologies that have been implemented in industrial and research based micromanipulators.

1.3 Gripper Manipulation Methods

In the world of micromanipulation, there exists 2 generic types of grippers. There are the active micro-grippers and the passive micro-grippers. For each type of gripper, there are advantages and disadvantages to them and should be chosen for a specific application. The details of how each category of grippers work are described below.

1.3.1 Active Micro-gripper

Active micro-grippers are grippers that actively open and close to grip objects willfully using an external means without being in contact with the object.

2

One of the most popular mechanism for actuation is done by running a current through a thermal actuator (asymmetric bimorph) to create a differential thermal expansion which creates an opening and closing action, as seen in Figure 1 below.

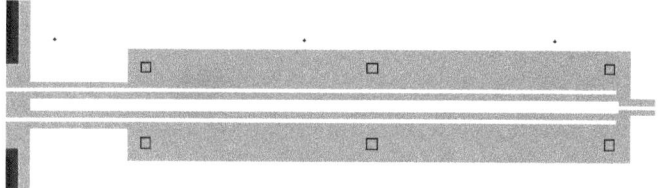

Figure 1: Two asymmetric bimorph thermal actuator in parallel to make a simple opening and closing movement [13]

The thermal actuated gripper can amplify its gripping strength by placing more thermal actuators in parallel to the gripping mechanism, as seen in Figure 2 below.

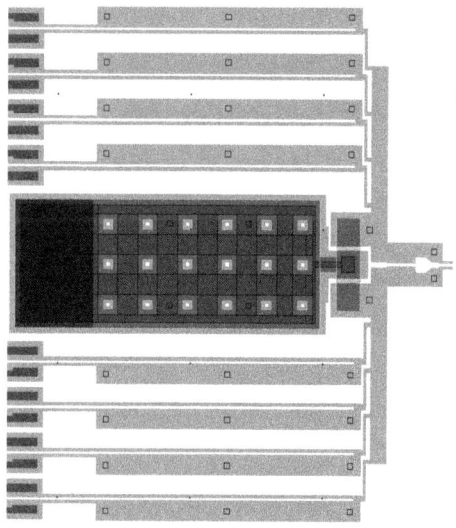

Figure 2: Amplified thermal actuated gripper [13]

3

Another very popular type of actuation used is the electrostatic actuation by means of a comb drive or parallel plate which has high voltage difference between two surface areas and this causes the two plates to attract. A pulsing voltage can create a vibrating structure or a means to actuate devices, as seen in Figure 3 below.

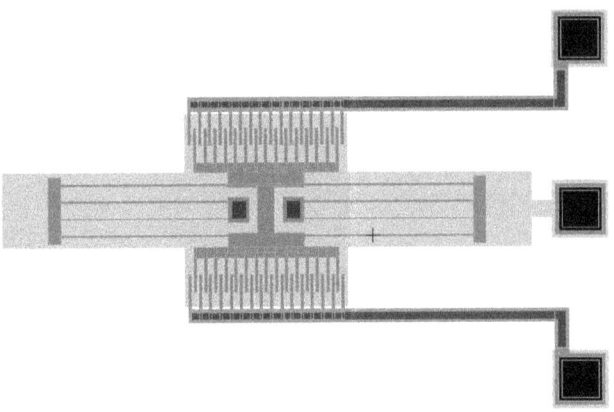

Figure 3: Electrostatic actuated devices using comb drives [13]

Other less popular but feasible means of gripper actuations are by means of piezoelectric actuation and magnetic actuators.

A well designed active micro-gripper is generally a very versatile end effector but also very expensive to fabricate and usually very bulky in design, which can occasionally cause it to warp or bend under its own weight.

1.3.2 Passive Micro-gripper

Passive micro-grippers make use of grippers that open and close by means of inducing an external force on the gripper tips. This is done by pushing the gripper tips against a fixed object and the micro-grippers open and close due to the elastic nature of the gripper beams, and therefore it is deem "passive", as seen in Figure 4 below.

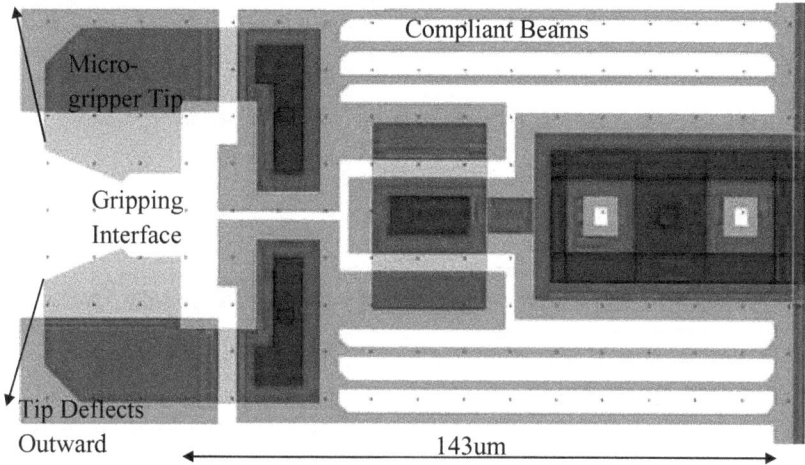

Figure 4: Passive Micro-gripper [5]

The main advantage of the passive micro-gripper is its simplicity in design and compact in size which also means that it is far cheaper to fabricate and implement than the active gripper. However, the passive micro-gripper would only be able to grab objects that are fixed in space by means of a mechanical tether that would be strong enough to allow the gripper to grip the micro-part but yet compliant enough to allow the micro-part to break off once the gripping is complete.

1.4 Micro Gripping and Joining Mechanisms

For the purpose of this project, we have chosen the passive gripper design based on a pre-existing passive gripper design to grip hold of micro-parts fixed in space by tethers. The reason for the passive gripper design is due to its proven track record of high reliability of successful gripping

5

and the low fabrication cost associated with it. The joining mechanism is a novel double compliant beam snatch and release slot that is designed for automation.

1.5 Computer Vision and Optics

The main feedback tool that will be implemented in this project is using computer vision. The main computer program used is Labview by National Instruments, as the main software for computer vision and robotic motion. Labview was chosen due to its versatile parallel programming capacity as well as its user friendly Visual Interface (.vi) programming platform.

In order for computer vision at a micro scale to be possible, it is required that vision tools to be in place to enable the user or computer to peer into the micro world. That means to choose the correct optics and image sensors for such purposes. There are a large array of optical microscope choices in the market, in this project, the confocal type microscope was chosen for practical and design reasons. The way it works is by flooding a light source that is reflected on a dichroic mirror onto an objective lens and onto the specimen. The reflected light rays from the specimen goes back through the objective lens past the dichroic mirror into a fluorescence barrier filter and through an aperture which is fed into the image sensor. The performance advantages of confocal microscopy as opposed to conventional optical microscopy is increased effective resolution with improved signal to noise ratio and also a clear depth perception in the Z-axis. All these attributes are imperative to the computer vision goals required in this project.

For digital processing, there exist 2 types of image sensors, they are the charged coupled device (CCD) and the complementary metal oxide semiconductor (CMOS) cameras. CCD based image sensors (cameras) have been known conventionally to produce a far better digital resolution as compared CMOS based cameras. However, as the technology of the latest CMOS cameras are rapidly improving to close the gap of digital resolution of CCD cameras, CMOS cameras offer a variety of advantages that CCD cameras do not. For example, CMOS cameras have much faster image processing speed and an adjustable region of interest (ROI) which CCD cameras do not have. The CMOS based image sensors were chosen in this project over the CCD based image sensors due to certain performance criteria requirements needed for real time micromanipulation.

The technique of recognizing the objects in the image space requires several post processing steps. The image needs to be first converted into grayscale, then the image would be further processed using filters such as the LoG (Laplacian of Gaussian) and then threshold for a certain intensity value before the Canny Edge Detection algorithm is applied to obtain the silhouette of the object. The object recognition algorithm utilized in the computer vision program uses an in-built Labview pattern recognition algorithm. The Labview pattern recognition algorithm uses a fuzzy logic form of statistical correlations between a template image and objects in the image space. If an object in the image space scores higher than a minimum statistical correlation to the template image, it would be recognized as an object.

The autofocus algorithm in the computer vision program is imperative to calculate contrast and bring the lens in focus to the specimen. What is commonly used out there in the industry is to use the highest contrast value in an image as a point of focus. However, as a form of novelty and also simplicity in the programming of the autofocus algorithm, the highest part score from the pattern matching algorithm in Labview was used as a measure for a part image to be in focus.

1.6 Motivation and Present Contributions

The need of efficient automated micromanipulation and assembly methods are required for large scale production of assembled micro-parts at a cheap cost. Computer vision is getting rapidly more advanced in the recent years and would serve as a reliable means to automate micro-assembly in the years to come. The goal of this work is to create an autonomous robotic manipulation system to provide an efficient and reproducible means of assembling micro-parts together on a chip. The intended academic contributions include:

1. Improving a 5 degree of freedom system to visually servo and manipulate the micro-gripper to assemble micro-parts together.

2. Utilize computer vision techniques to properly recognize and determine a micro-part's position in space and successfully join and assemble them by instructing the motion controller the exact distance and trajectory it needs to complete the move

3. Developing better micro-part designs with higher assembling success rates.

4. Developing a micro-assembly specific MEMS chip and streamlining the assembly process by considering the part logistics and the assembly pipeline.

The proposed system is intended to provide a fully autonomous solution to micro-part assembly. This includes micro-part identification using pattern recognition algorithms from computer vision followed by micro-part assembly.

1.7 Dissertation Outline

This thesis contains 10 chapters. Which are outlined in the order described below.

In Chapter 1, the problem statement of the project was defined followed by a survey of the state-of-the-art micromanipulation techniques specifically on MEMS micro-parts and manipulation techniques. Chapter 2 discusses the system methodology and overview. In Chapter 3, it describes the current chip micro-part design and modifications to older designs as well as the streamlined logistic process of the micro-assembly to increase the probability of micro-

8

assembly success rate as well as the speed of assembly. Chapter 4 presents the current 5 DOF used to assemble MEMS micro-parts and another more precise and accurate 5 DOF system that is in its midst of development. Chapter 5 describes the computer vision process of acquiring images from the camera and using image post processing techniques and pattern recognition to recognize the position of micro-parts in space and assembling them together. Chapter 6 explores the concept of using a novel set of visual markers on the chip to calibrate the chip's spatial orientation, and set a spatial origin which aids the visual servoing of the autonomous system. Chapter 7 describes in detail the automation process of using the computer vision synchronously with the motion control and the specifics of the programming procedures with reference to the mechanical processes. Chapter 8 summarizes the conclusion and contributions of this project. Chapter 9 discusses recommendations and the suggested future research directions, and finally Chapter 10 concludes the report.

Chapter 2

Methodology and System Overview

2.1 Micromanipulation System

The goal of the MEMS micromanipulation system is to create a robotic system to assemble micro-parts together in a 3D (three-dimensional) format. To work towards a completely autonomous system, means that the user must set the variables of the computer vision requirements, and must specify the items to be micro-assembled, to let the computer run autonomously to achieve the micro-assembly results without any manual intervention from the user.

To achieve such a feat, several subsystems have to be in place including: a multi-degree-of freedom electromechanical system attached to a rotatable end-effector for micromanipulation. In addition to that, an optical subsystem integrated with a computer vision system as a feedback tool for the motion controller to work from.

Combining these subsystems together, the micro-assembly process for the MEMS micromanipulator works as follows:

1. Scan the MEMS chip for visual markers (fixed reference points on the chip)

2. Calculate the degree of rotation offset the chip is from the fixed reference point and re-orientate the chip correctly

3. Re-servo the chip for visual markers and find the referenced zero marker. All the relative X, Y and Z ordinates are re-calibrated as zero.

4. Micro-assembly is ready to begin. The user or computer searches for the closest available micro-gripper.

5. The user attaches the micro-gripper to the end-effector using a UV-curable adhesive. A slight rotation of 3.6 degrees into the plane is achieved. (This process still cannot be automated due to physical constraints)

6. Visual servoing to the closest available micro-part.

7. Using computer vision and automated gripping, the micro-gripper grabs hold of the micro-part and peels it out of its tethers.

8. The micro-part is re-oriented 90 degrees to the plane of the chip.

9. Micro-part is joined manually to a slot.

Chapter 2 gives the reader an overview of the system mechanics starting with the visual servoing task of the MEMS micromanipulator followed by the actual physical manipulation of the micro-parts. Description of each subsystem will be mentioned at the end of each task and referenced to the later chapters.

2.1.1 Overview of Visual Servoing

To set the optical coordinate system on the MEMS chip, several fixed reference visual markers were placed on the chip surface to guide the micromanipulator to the correct position in space. The visual markers are placed all across the MEMS chip as shown in Figure 5 below.

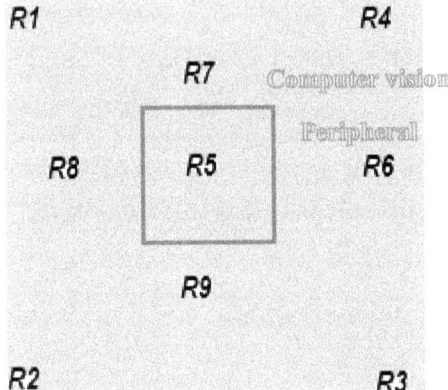

Figure 5: Chip surface with reference points (R1, R2...) laid out across the chip substrate

The use of multiple visual markers across the chip is required to overcome the limitation of having a fairly small computer vision peripheral. In this case, the MEMS manipulator is able to randomly choose any visual marker on the chip and immediately know where it is exactly on the MEMS chip. The actual visual markers are in the form of circles and not the alphabet "R", but for a conceptual understanding, they are represented as R1, R2, etc... More in depth discussion on visual markers are discussed in Chapter 6. An example of an actual visual marker is seen in Figure 6.

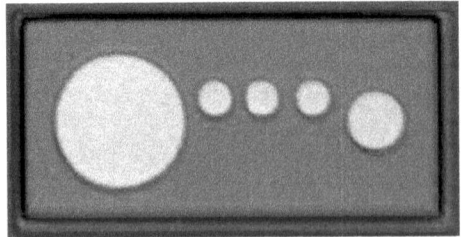

Figure 6: An example of a circular visual marker used on the chip – R3 (3 small circles)

12

Once the MEMS manipulator has established where it is in space, and corrected for spatial misalignments, it is able to determine where the parts are on the chip, as seen in Figure 7.

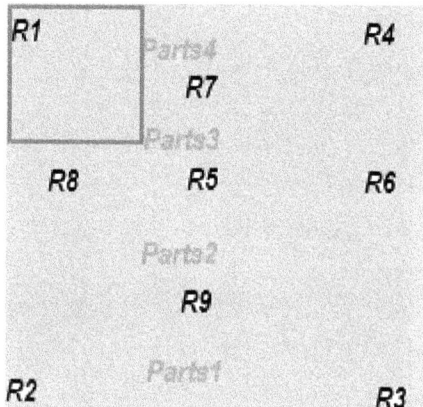

Figure 7: Once the computer vision has established the referenced points, it knows where all the parts are with relation to it.

2.1.2 Overview of Micromanipulation Using Tele-operation

The first step to micromanipulation requires the end-effector to be attached to a micro-gripper. This requires the MEMS manipulator to visually servo to the spot on the chip where it expects to have a micro-gripper and verifies that the micro-gripper is indeed there. Otherwise, it will move on to the next available micro-gripper. Once the micro-gripper has been verified to be on the chip the user will adhere the micro-gripper to the end-effector via a UV-curable adhesive, as seen in Figure 8.

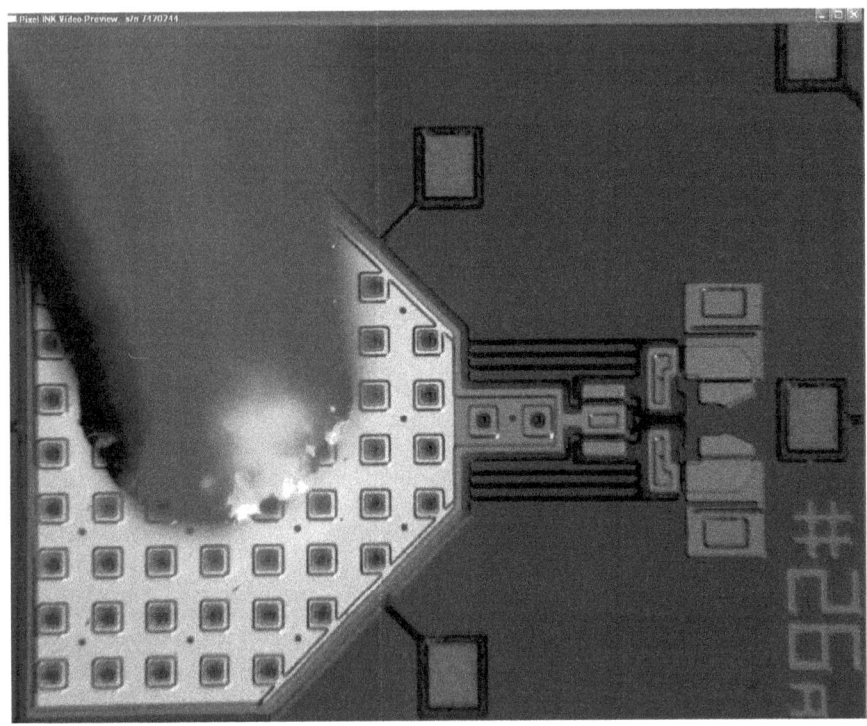

Figure 8: Micro-gripper adhered to the end effector using UV-curable adhesive (The adhesive is undergoing UV-curing)

Once the micro-gripper is adhered to the end-effector, it is pulled out of its anchor and rotated 3.6 degrees into the plane of the chip and the user searches for the nearest micro-part to grab on. Once the micro-gripper and micro-part are on the same image space, the user initiates auto-gripping of the two parts. Ideally, when the micro-gripper is successfully bonded to the end-effector, the computer will completely automate the searching process and initiate the gripping of parts. However, several subsystems are not yet in place for that process to be fully functional, but it is a close reality. The micro-gripper pulled from its anchor, shown in Figure 9.

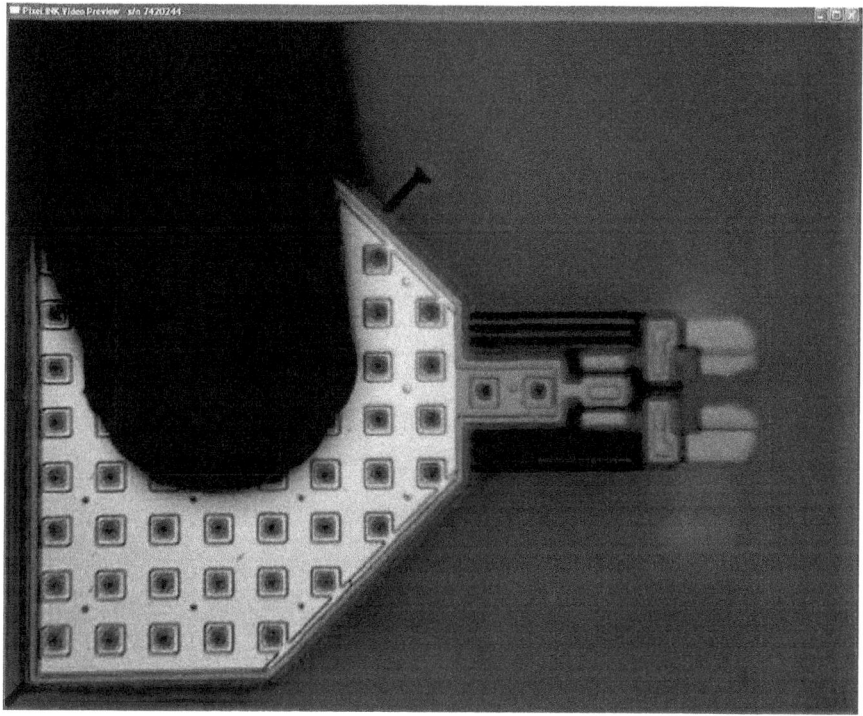

Figure 9: Micro-gripper is pulled out from its anchor and rotated 3.6 degrees into the plane of the chip

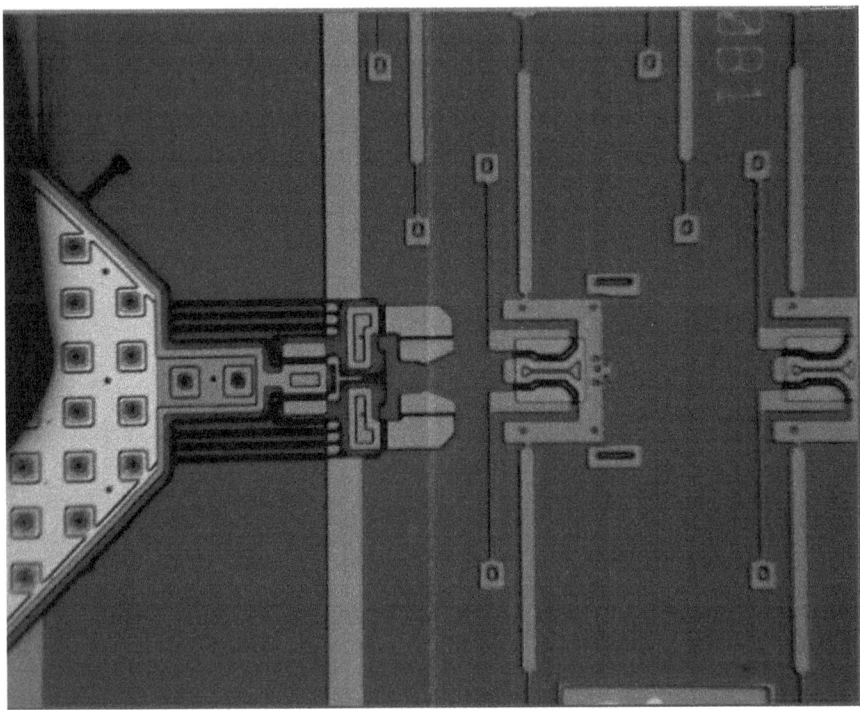

Figure 10: User or computer searches for the nearest micro-part on the chip for joining

In Figure 10, the way the computer automates the joining process is by identifying the micro-gripper and micro-part via pattern matching of pre-define templates. Once the two parts are identified on the image space, they are simply joined together by measuring the distances between them. The computer does not naturally know how far the parts are away from each other, however, it does know how far the parts are from each other by counting the pixels between them. The pixel distances are converted into distances measured in microns in a subroutine. The computer then does a final conversion to measure how many stepper/servo rotations is required to complete the motion, as seen in Figure 11.

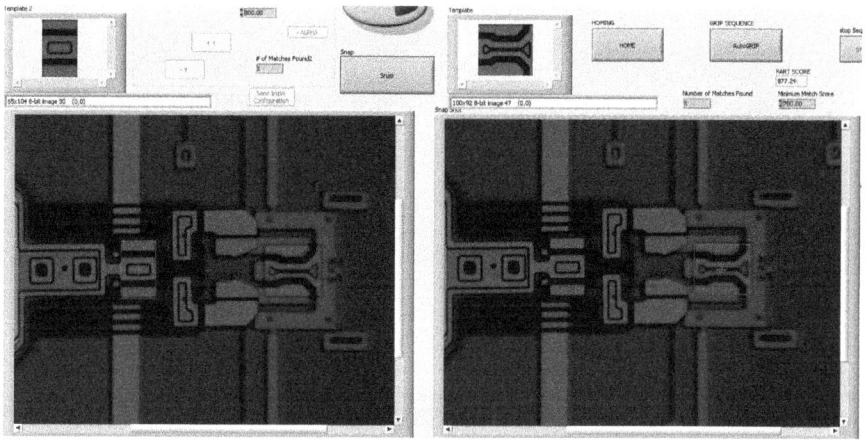

Figure 11: Computer vision recognizes the 2 distinct parts in the image space and calculates the x and y distances between them

The joining process is more complicated than it is shown in the figures. Although depth of field of the optical system is 1.5 microns, one can assume that if both the micro-gripper and micro-part are in focus, they can be estimated to be fairly close to being on the same plane. However, the gripping process is not as simple as making both images be in focus and joining them. It requires a series of iterative processes of trying to grip the micro-part and backing away to move a single micron down in the Z-axis and trying it again until gripping is successful. A more detailed description of this process will be discussed in Chapter 7 on the automation procedure. Figures 12 and 13 will briefly illustrate this process.

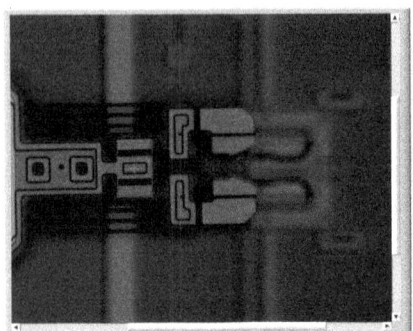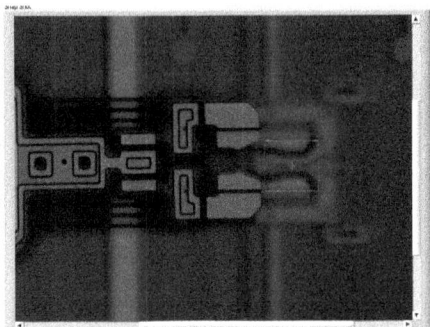

Figure 12: Usually the micro-gripper and micro-part is out of plane from one another, the computer vision process has to also account for the offset in the Z-axis

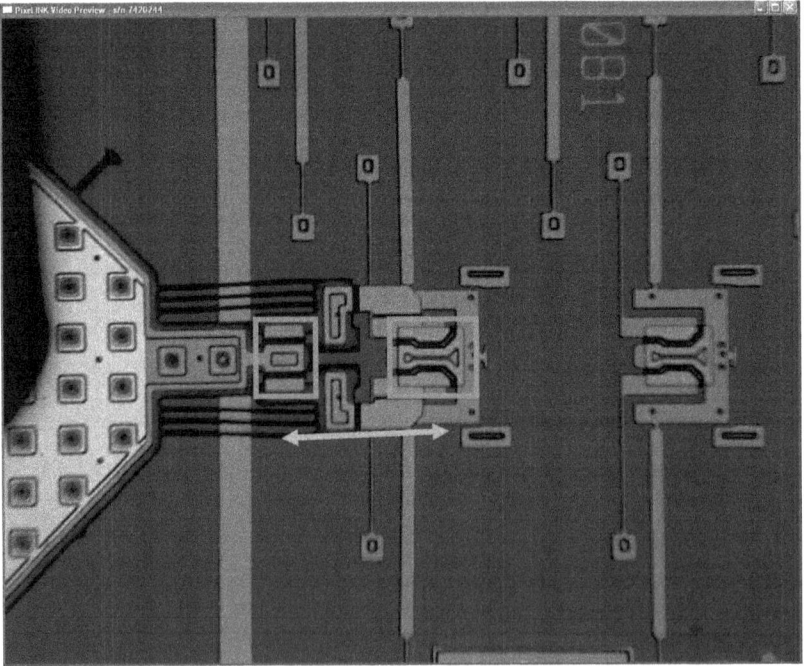

Figure 13: The computer measures the distances between the two parts to determine whether the part is joined

Once the distances between the two parts get lower than a certain value, we can assume that the micro-part is grabbed on by the micro-gripper and ready to break off. The computer initiates a finishing routine that pushes the micro-gripper in the X-axis until the micro-part tethers break and the micro-part is ready to be lifted off the chip substrate. This process is shown in Figures 14 and 15.

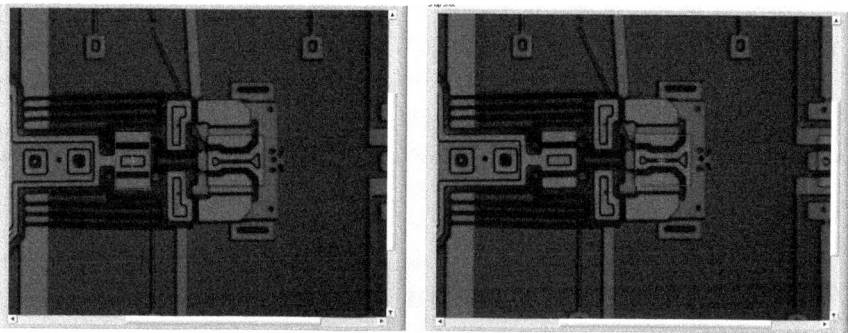

Figure 14: Micro-part is broken off from its tethers

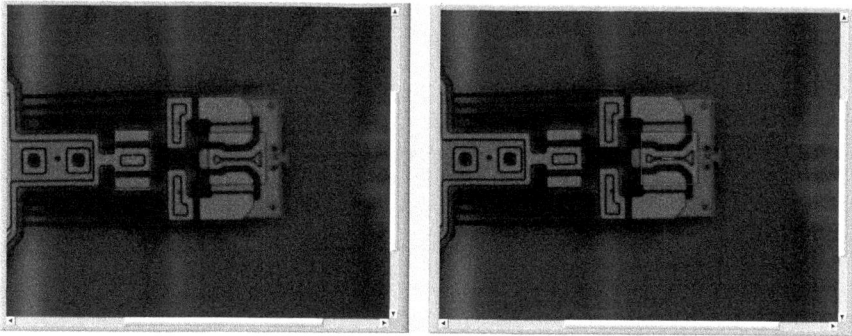

Figure 15: The micro-part is lifted off the chip substrate

A more in depth discussion on the computer vision process working simultaneously with the motion control for the automation process will be discussed in Chapter 7.

2.1.3 Micro-joining

Once the micro-part is grabbed by the gripper and pulled out of its spot, the MEMS manipulator will search for the micro-part slot to be joined. This requires the end-effector to first rotate a full 90 degrees so that the micro-part is perpendicular to the chip substrate. From there, the micro-part can be pushed into the micro-part slot and erected in a vertical fashion. Ultimately, being able to manipulate micro-parts in a three dimensional fashion is the crux of this research. Conventional MEMS technology builds devices from the ground up using photolithography. What micromanipulation does is giving MEMS technology an additional degree-of-freedom of assembling parts off the plane of the chip substrate, which will arise many more applications such as micro-mirrors, beam splitters etc... The process of joining is shown below in Figure 16.

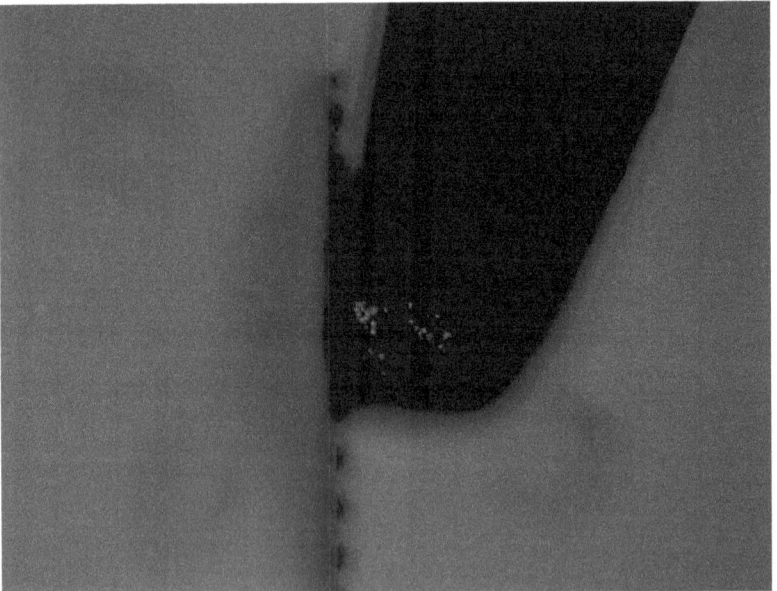

Figure 16: The end effector is rotated 90 degrees to the plane of the chip

Figure 17: The user searches for a slot for the parts to slide in (The shadow is caused by the end effector)

In Figure 17, once the micro-part is aligned above the slot to be joined, the user then refocuses the camera onto the tip of the micro-part as seen in Figure 18.

Figure 18: The microscope is refocused on the tip of the micro-part

21

The micro-part is lowered slowly down towards the chip substrate. As the part gets closer to the chip substrate, the compliant beam slots starts to get to be in focus. When the micro-part physically touches the chip substrate, it deflects very slightly. At this point, the micro-part is verified to be in contact with the slot and is ready for insertion, as seen in Figure 19.

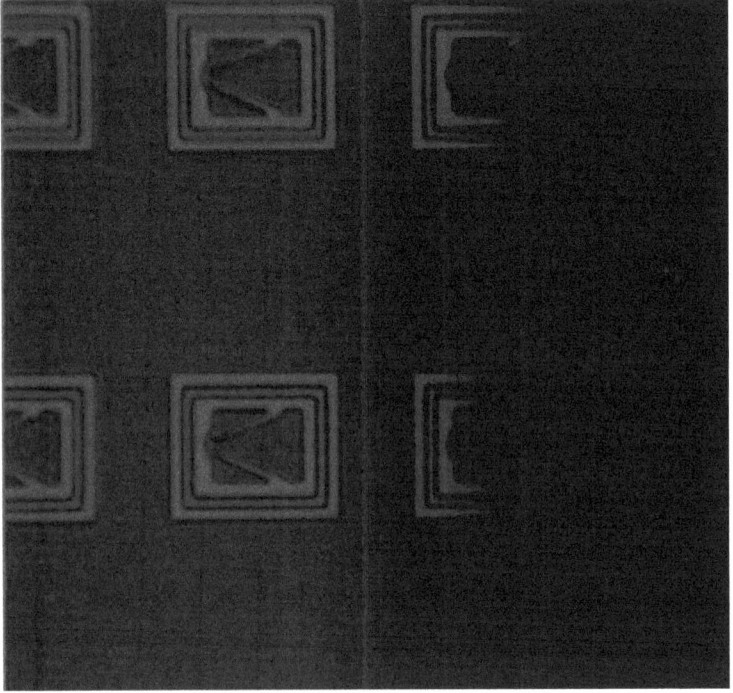

Figure 19: The micro-part is lowered to the plane of the double compliant beam slot

As the micro-part slides across the double compliant beam slot, at various points on the sliding process, external perturbation by means of vibration is required to ensure that the micro-part does not stick to the compliant beams. External perturbation in this particular example is created by lightly tapping the system, as seen in Figure 20.

Figure 20: The micro-part slides across the double compliant beam slot until it reaches the end of its travel

Once the micro-part is successfully joined to the slot, the end effector is raised from the surface of the substrate. This causes the micro-gripper to release the micro-part and the micro-part stays connected to the joining slot, as seen in Figure 21.

Figure 21: The micro-gripper is raised above the chip substrate, releasing the micro-part as it stays connected to the joining slot

This portion of the micro-assembly concludes the first successful cycle of the micro-assembly process, from which the process can be re-iterated indefinitely until the final product is assembled.

The details of the automation principles and programming principles will be discussed in detail in Chapter 7.

Chapter 3

Micro-part Design and Chip Layout

3.1 Micro-part Designs

This chapter focuses on the philosophy behind the microchip parts design and mechanisms used which is integrated to the use of micromanipulation. It will discuss in depth the several micromanipulation mechanism strategies used in current research, the part design improvements done on existing micro-parts to make it more reliable, visual markers used to aid the alignment of the chip and serve as a reference point in space and last but not least, to streamline the assembly process the parts are placed in a logistical manner to reduce the time required for effective assembly.

3.1.1 Micro-gripper

The micro-gripper utilizes the "passive" gripping technology, which means that it uses the elastic spring effect of the silicon beams to bend back and forth to grip onto objects. The micro-gripper has to push itself against a micro-part fixed in space before its grippers would open up due to an external force and grip the object. To do so, the micro-parts have to be fixed in space using sacrificial tethers in which are meant to break off when the micro-grippers have successfully gripped the object and pushed it out of place. A passive micro-gripper was chosen over an active one because of cost effectiveness of fabrication, simplicity of design and ease of implementation.

The gripper dimensions and stresses were simulated on SolidWorks to optimize its existing design parameters, as seen in Figure 22.

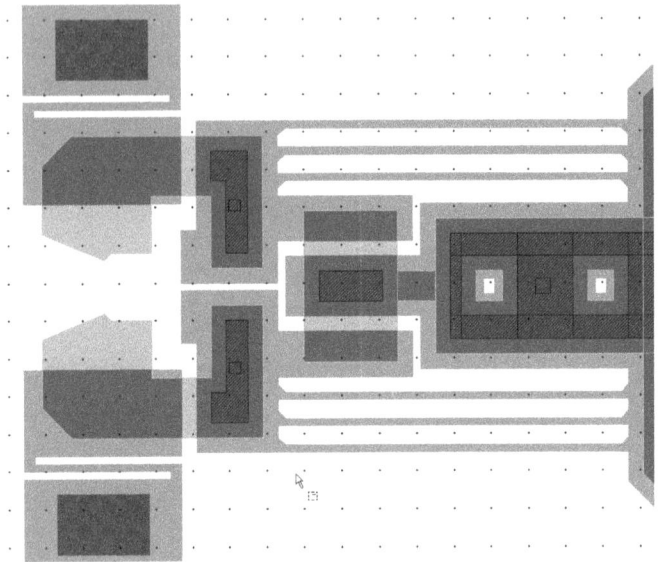

Figure 22: Original Micro-gripper Design in L-Edit [5]

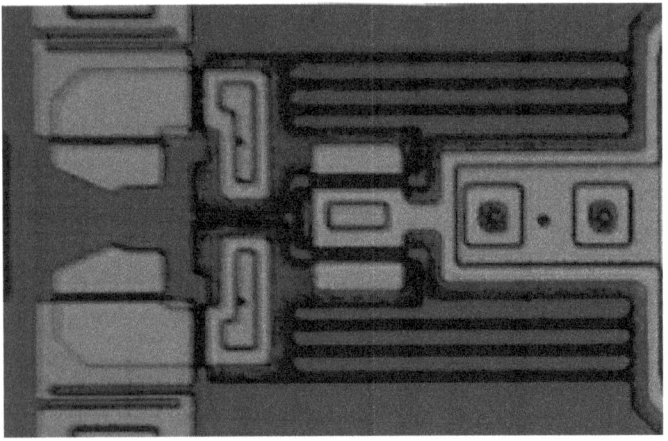

Figure 23: Fabricated Micro-gripper

A separate kind of passive micro-gripper was also fabricated on this chip. It utilizes the same principles as the micro-gripper described above, but separates the grips widely. This helps to reduce the turning moment of the wide micro-part being gripped and the deflection of the micro-part being held, as seen in Figure 24.

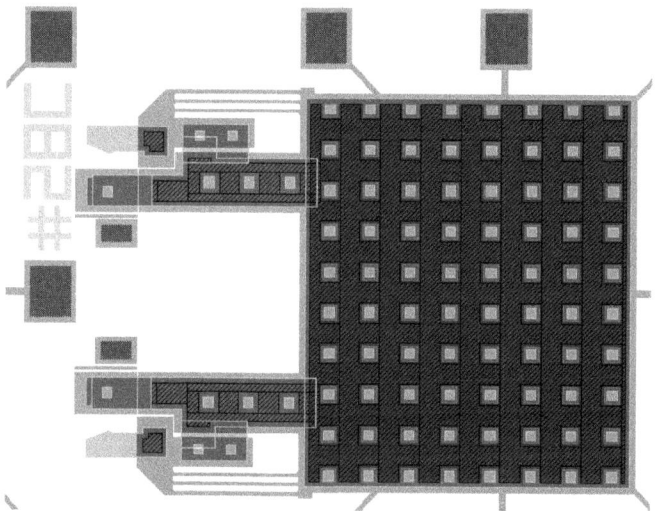

Figure 24: L-Edit design of wide micro-gripper [5]

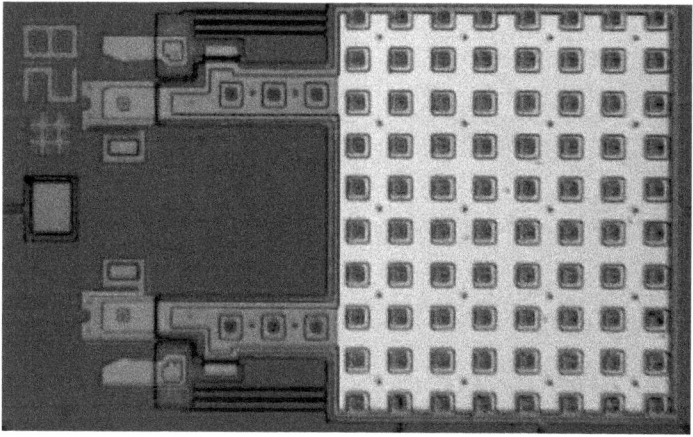

Figure 25: Actual fabricated micro-gripper

27

3.1.2 Micro-part

The micro-part designs range between 80um – 150um. The designs are made in a modular way to allow consistency and simplicity in construction and ease of replacement if it breaks. The idea behind the micro-parts is to have a mating gripping interface with the micro-gripper and a joining interface with the chip substrate via a quick snap and release mechanism. Improvements on this existing part includes smaller tolerances on the guide to prevent part rotation and premature breaking of micro-part and longer modified tethers to concentrate the stress forces on the joints for clean break-off, as seen as Figure 26.

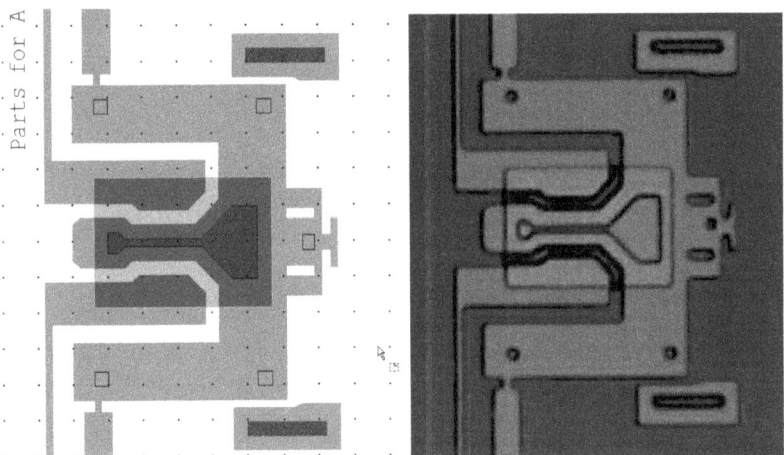

Figure 26: Micro-part design on L-Edit on the left and the actual fabricated part on the right [5]

The forces of the tethers was simulated on SolidWorks to see where the stress concentrations exist and where it would most likely break when a force is exerted.

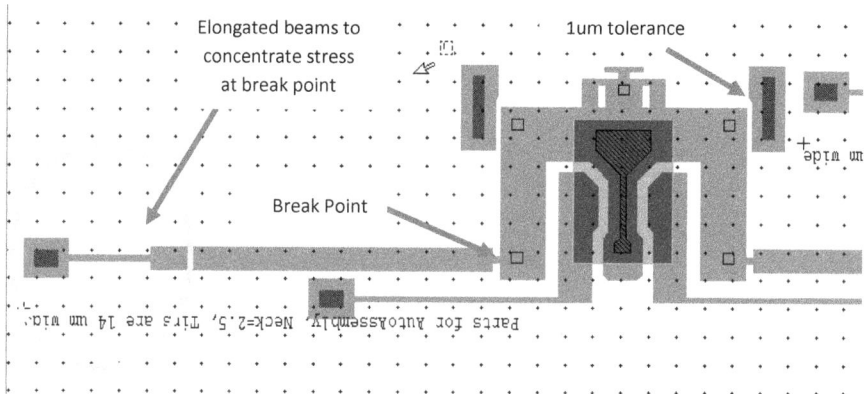

Figure 27: Elongated beams help concentrate the stress at the break points

In Figure 27, the reason why the beams are elongated close to the anchor point is such that when the micro-gripper pushes the micro-part to break away, the tethers would bend much more at its ends and help concentrate the stress forces on the break point. This facilitates for a much cleaner break away than previous designs. The one micron tolerance on the guides also help prevent the micro-part from rotating when it is being pushed out, making it an overall more cleaner and more successful break away design.

Similar design changes were also implemented on the wide micro-parts that utilizes the micro-gripper for manipulation. The wide micro-parts also underwent design changes, it has slots on the top, bottom as well as the side of it. This makes the wide micro-parts modular and interchangeable in its arrangement of assembly, as seen in Figure 28 and 30.

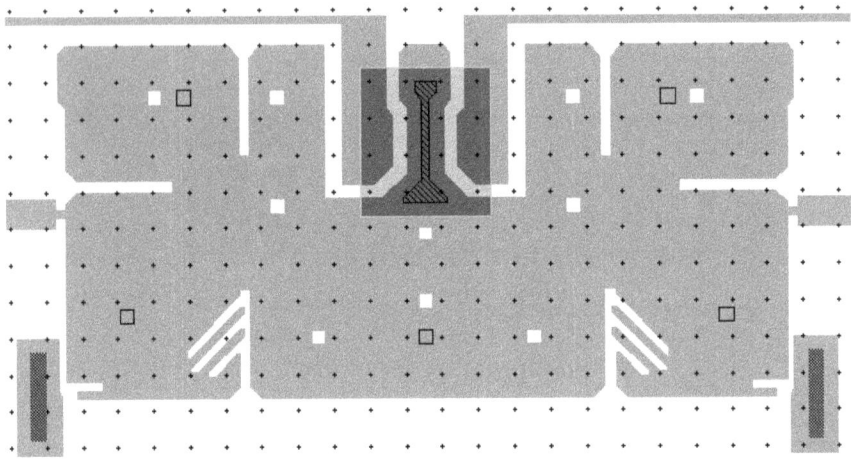

Figure 28: L-Edit design of the wide micro-part that utilizes the narrow micro-gripper

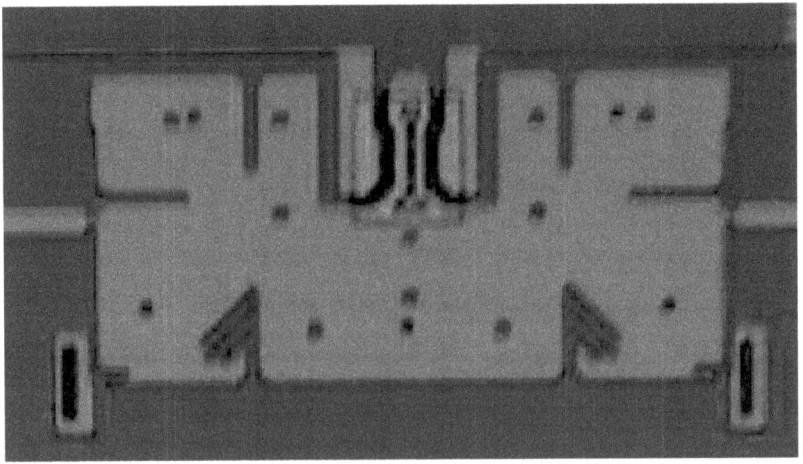

Figure 29: Actual fabricated wide micro-part

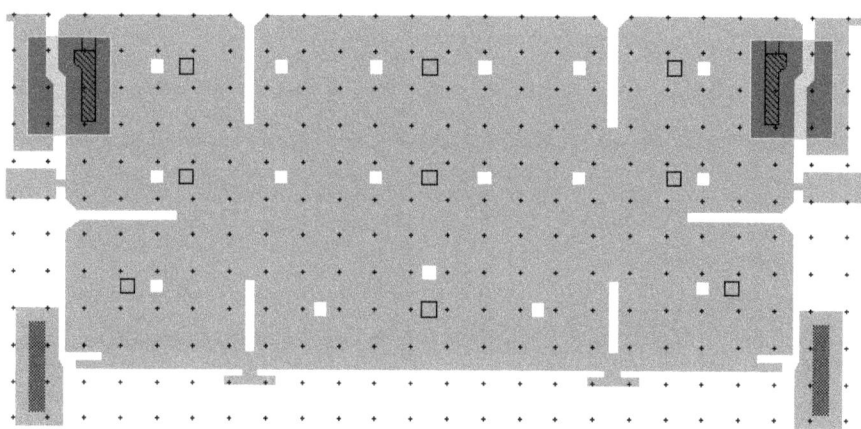

Figure 30: L-edit design on the wide micro-gripper that utilizes the wide micro-gripper

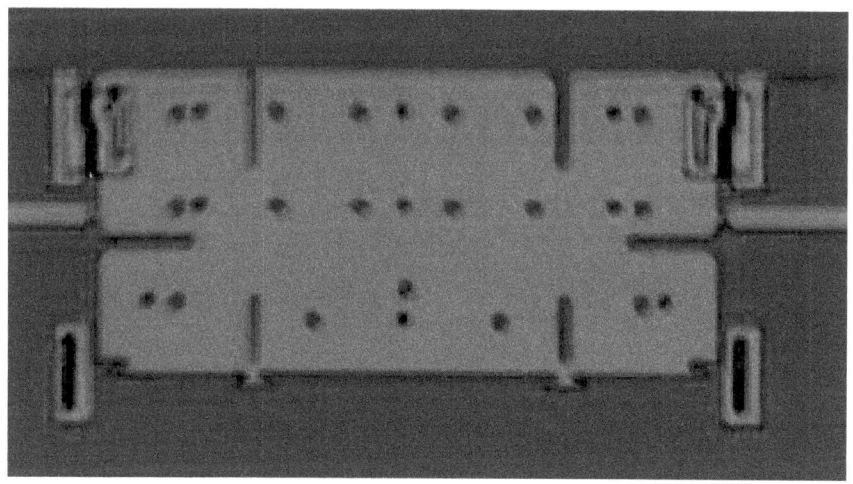

Figure 31: Actual fabricated wide gripper

3.1.3 Joining Slots

The older joining slots uses a T-profile slot with a single compliant beam, which the micro-parts would slip in place. The compliant beam serves as a form of snap release forcing the part to be fully seated in the slot when it has moved to the end of its travel. However, the tolerances and dimensions of the T-slots were not very well optimized and made the pre-existing joining process very cumbersome and unreliable, as seen in Figure 32. The newer more optimized modified joining slots uses a double compliant beam profile instead of one. This not only restricts horizontal movement and fixates the micro-part in space in a much better configuration, it also allows more play before the part has reached the end of its travel. The large entry slot of the new joining slot is to ensure that the joining success rate would be higher and much easier for the user and computer to achieve, as seen in Figure 33 and 34.

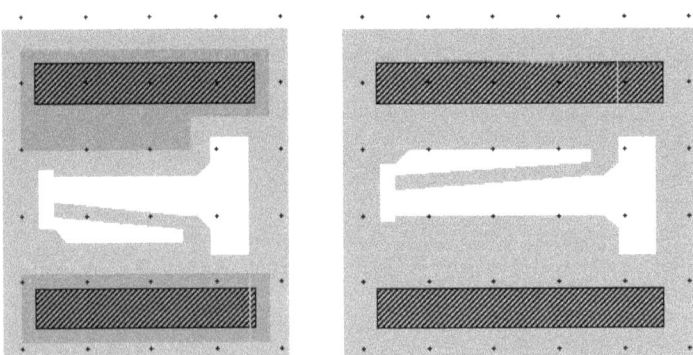

Figure 32: The two generic older type T-slots that were used for joining [6]

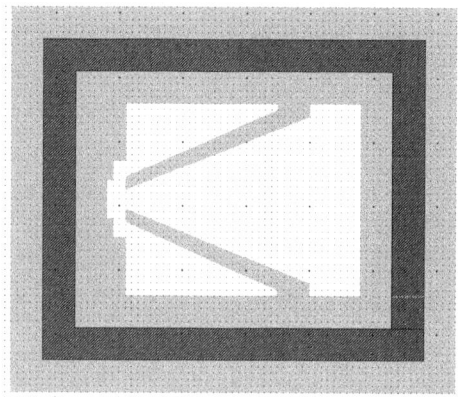

Figure 33: Double compliant beam slot that is currently used for joining

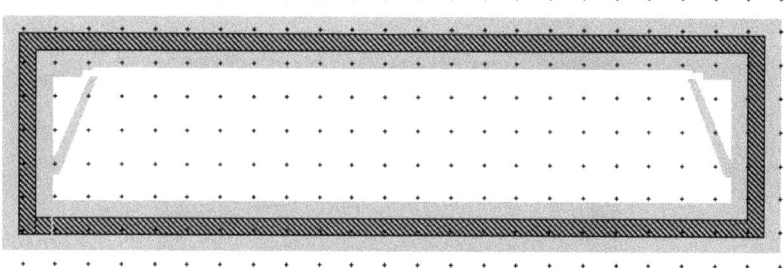

Figure 34: Double compliant beam slots for wide parts

3.1.4 Visual Markers

Visual markers are implemented in this chip design in order to aid the computer vision process. What these visual markers do is to give orientation information and reference point information to the computer and user. The visual markers are made of three different types of circles, the large, medium and small circles. The large and medium circles determines the orientation information of the chip and the small circles determine which visual marker this refers to and where it is on the chip, as seen in Figure 35.

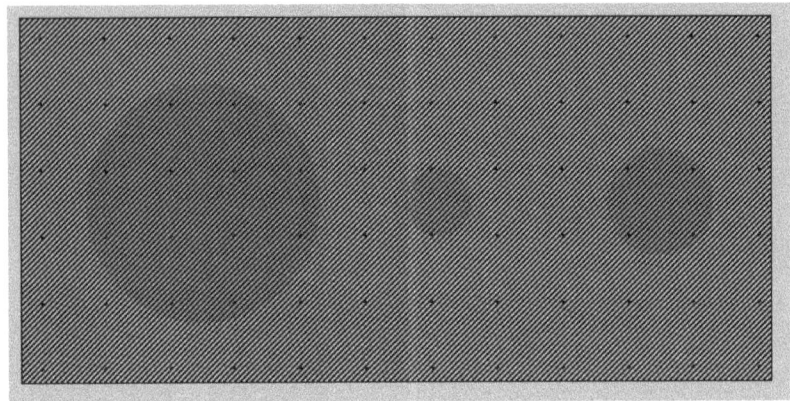

Figure 35: Visual Markers. Large, medium and small circles. Metal on top of Anchor2 which is laid on top of Poly2

Figure 36: Actual fabricated marker

These types of visual markers are a fairly novel concept and quite an extensive background to how they are fabricated and utilized in the computer vision process has to be explained in detail. Chapter 6 is dedicated solely to explaining how they work.

3.2 Micro-part Layout

The micro-parts are laid out in a parallel configuration to reduce time required for the motion controller to re-orientate and locate a part. It also streamlines the gripping and joining process by minimizing the number of steps required to achieve a successful assembly. This streamlining of the assembly process is crucial for reducing unnecessary movements and have a quicker part assembly turnaround time. The layout of the chip is shown in Figure 37 below.

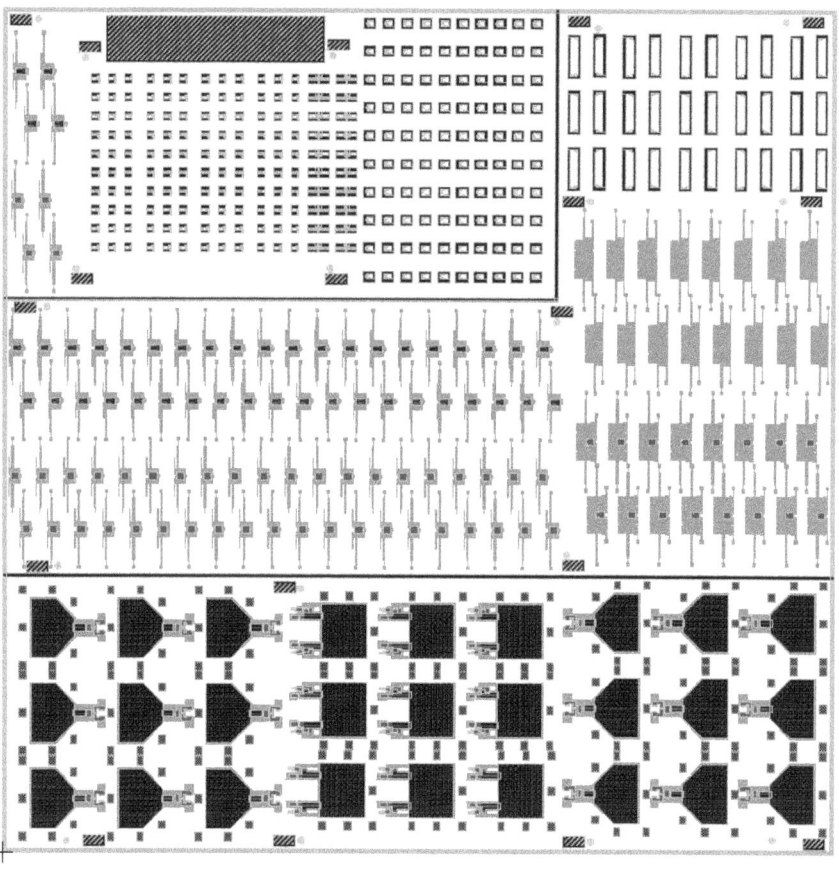

Figure 37: L-Edit layout of the chip

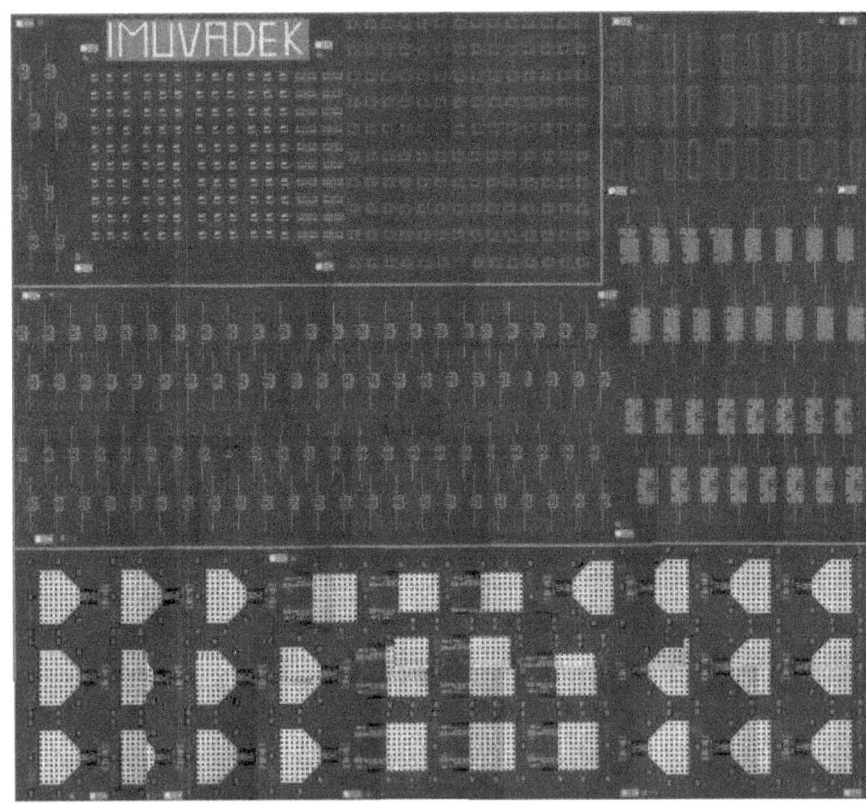

Figure 38: Actual Part fabrication - stitched image

Chapter 4

5 Degree of Freedom Robotic System

4.1 Current Robotic System

The current robotic system in use, as shown in Figure 39, uses a 5 axis stepper motor system to maneuver and manipulate the micro-parts. It also uses Galil controllers and Galil programming language to control the stepper motors. However, the user interface and the main programming language used to execute the commands is done in Labview Virtual Instruments (.vi) and the manipulator system has approximately 1 micron step resolution.

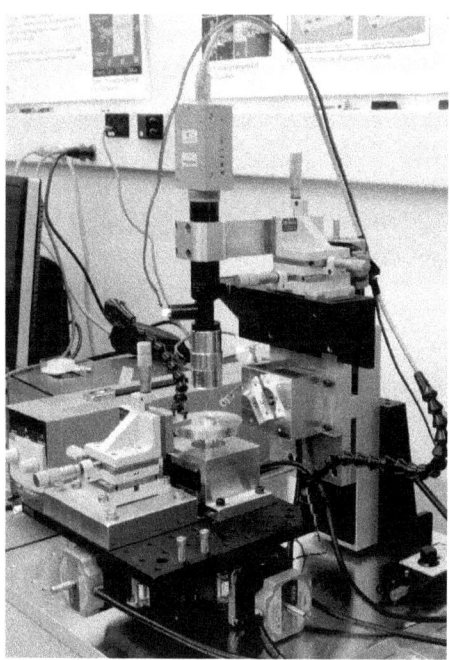

Figure 39: Current 5-DOF robotic system [4]

Figure 40: Drivers and Galil DMC-2143

Figure 40 above shows the Galil DMC-2143 controller (black box) that gets command information

from the computer via an Ethernet cable and converts the Galil commands into pulse signals which

are fed into 5 separate drivers. These drivers amplify the signal and send a five phase signal to the

respective stepper motors. Figure 41 shows the entire micro-assembly work station set up.

Figure 41: Robotic system and computer work station

The current robotic system is robust and reliable enough to enable automated micro-gripping. However, the optical train is currently held by a manual stage and requires the user to adjust it to regain focus on the chip parts at various parts of the process. This undermines the ultimate goal of a fully automated system. Stepper or servo motors have to put in place of the manual stages before it can be programmed to be fully automated.

4.2 Developing Robotic System

The ongoing development of a robotic system uses a 4 axis servo quadrature encoder with a high resolution stepper motor actuated end-effector. This robotic system provides a much higher and more precise movement resolution as compared to the current existing system and would be more ideal to use it for smaller parts and automation purposes. It also has the optical train actuated by 3 axis stepper motors, which takes out the need for the user to manually adjust the stages to get focus on the specimen, as seen in Figure 42.

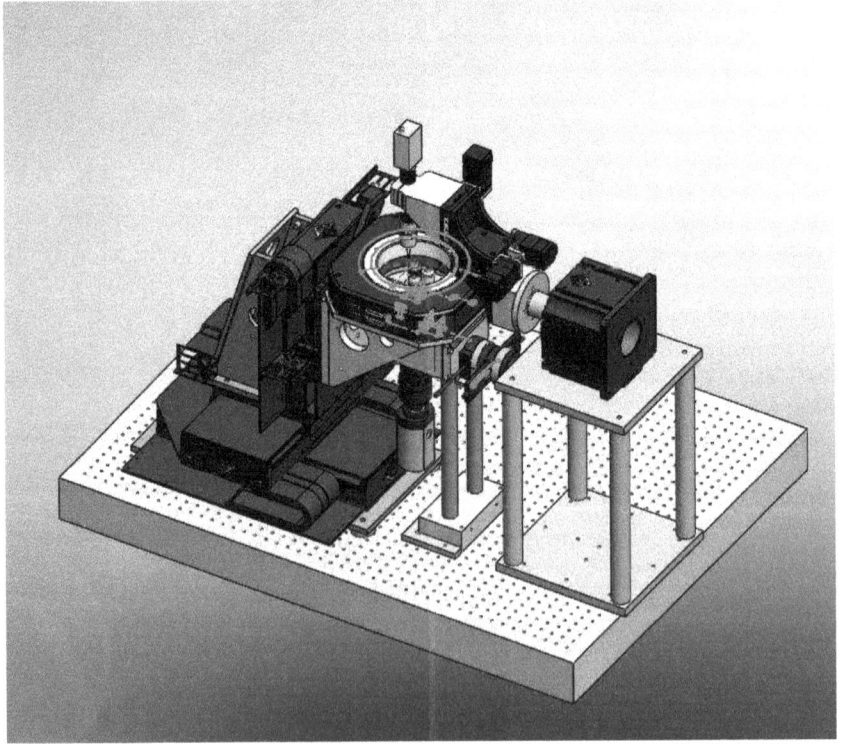

Figure 42: Developing 5DOF machine utilizing Aerotech servo motors

Chapter 5

Optical Train and Computer Vision

5.1 Optical Train

Resolving detail in MEMS imaging requires sophisticated hardware and optical lenses. Conventional microscope lenses have a tendency to have a fisheye effect on the edges of the image space. This distorts the actual object of interest and would not be ideal for machine vision purposes. The type of microscope objective lens utilized is a Mitutoyo - Edmund Optics Infinity-Corrected long working distance objectives (HR 20x/060, ∞/0, f = 200). The optics creates a plane field of view, which greatly reduces the fisheye effect and keeps the images rectilinear and free from distortion for the most part of the image space. The long working distance objective allows the end-effector to work simultaneously in between the optics and the chip and rotate and manipulate parts without any obstruction or difficulty, as shown in the Figure 43 below.

Figure 43: Microscope and end-effector working simultaneously

The confocal type microscope was chosen for practical reasons. The way it works is by flooding a light source that is reflected on a dichroic mirror onto an objective lens and onto the specimen. The reflected light rays on the specimen goes back through the objective lens past the dichroic mirror and into a fluorescence barrier filter and through an aperture which is fed into the image sensor. The performance advantages of confocal microscopy as opposed to conventional optical microscopy is increased effective resolution with improved signal to noise ratio and also a clear depth perception in the Z-axis. All these attributes are imperative to the computer vision goals required in this project.

5.2 Image Sensor (Camera)

The image captured by the microscope's optics are digitized using a complimentary metal-oxide semiconductor (CMOS) image sensor which is capable of providing a relatively high number of frames-per-second in comparison to its rivaled technology, a charged coupled device (CCD) image sensor. The particular CMOS camera used is a PixeLink PL-A742. It has a maximum digital resolution of 1280 X 1024 at 27 frames per second. The high resolution and high frame rate of the image sensor makes it an ideal candidate for this application.

Current digital imaging technology is produced commonly in two forms: CCD (Charged Coupled Devices) and CMOS (Complementary Metal-Oxide Semiconductor) sensors. CCD cameras are known conventionally to have superior image quality as compared to CMOS based cameras. However, in the recent years, the cost and quality of CMOS cameras have gradually caught up with the CCD based cameras.

CCD image sensors can be thought of as millions of arrays of microscopic wells, known as photosites, which are metal-oxide semiconductors that collect photons of light and convert them to electrical signals. The sensor array collects a sample of light where each row of photosites passes the collected charge to the next row closer to the edge of the array. Once the charge is passed to the row that resides at the edge of the array, the charge from each photosite is passed to a voltage converter and amplified accordingly. Although the image captured is of high quality it comes with an expense, this process is sequential and cumbersome, causing the entire procedure to suffer from a low frame count and high processing latency. CCD based cameras are conventionally used in astronomy for their high resolution, low noise and high quality of image captures, but would not be ideal for high speed cameras and hand held devices.

CMOS image sensors are a constantly evolving technology and the quality of image captures is improving at such a rapid pace that it is approaching that of the CCD based image sensors. Although the image quality of CMOS cameras have not yet completely out phased the CCD based image sensors, there are compelling characteristics of the CMOS image sensors that makes it a good choice over the CCD image sensors for the purpose of computer vision and MEMS micromanipulation. The CMOS grid unlike the CCD sensor consists of integrated photosite commonly including a voltage converter, dedicated select and reset circuitry and an onsite amplifier. The on-site dedicated components increase the functionality and the speed of the sensor, as it does not bottle neck the image processing sequence like the CCD based technology does. In the CMOS technology, the voltage sensed by individual grids do not have to be passed down to adjacent grids in a sequential manner, before the voltage signals are processed, making it acquire a far faster and more efficient processing speed.

However, the downside of the CMOS image sensor is that it reduces the overall ratio of light collecting area per site, thus decreasing its digital resolution. The primary advantage of CMOS sensors is not only speed, but also the ability to dynamically select a region-of-interest

(ROI), or a subsection of the entire image space, allowing a reduction in pixel-processing latency without a loss in image window resolution. This is highly advantageous for post-processing images in real-time applications such as computer vision and real time MEMS micromanipulation. ROIs can be reduced to focus only on the micro-gripper and the micro-part, ignoring all other components in the image space. This improves image processing speed as well as noise reduction in the image space. In addition to performance gains, CMOS sensors are produced at a fraction of the cost of CCD sensors, which makes it a more cost effective solution.

Overall, the CMOS sensor was selected over the CCD sensor for its ability to capture a high frame rate, select regions-of-interests within the image space and the low image processing latency which is required for real time computer vision and micromanipulation.

5.3 Camera Exposure

All the images read out by the sensor from the computer will output a raw RGB (Red Green Blue) image by means of a Bayer filter interpolation of individual RGB grids. However, the quality of the image is also highly dependent on the Exposure Triangle, which is ISO, Shutter Speed and Aperture. The shutter speed and ISO values can be adjusted in the software configuration of the image sensor, but the aperture size is fixed in the hardware.

ISO is the level of sensitivity the camera is to available light. It is typically measured in numbers, a lower number representing lower sensitivity to light and a higher number means higher sensitivity. More sensitivity comes at a cost of higher noise or grain in the image. (E.g. ISO: 100, 200, 400, 800, 1600)

Shutter speed is the length of time a camera shutter is open to expose light into the camera sensor. Shutter speeds are typically measured in fractions of a second. Slow shutter speeds allow more light into the camera, which is typically used for low-light or night photography, while a fast shutter speed capture objects moving at a high speed in high light settings. (Examples of shutter speeds: 1/1, 1/15, 1/30, 1/60, 1/125 etc…)

Aperture refers to the hole within a lens which light travels through into the camera's image sensor. The larger the hole, the more photons is able to pass through to reach the sensor. Aperture also controls the depth of field, which is the portion of a scene that appears to be sharp. If the aperture is small, the depth of field is large and if the aperture is large the depth of field is small. In photography, the aperture is typically express in "f" numbers (known as "focal ratio", since the f-number is the ratio of the diameter of the lens aperture to the length of the lens. Examples of f-numbers are: f/1.4, f/2.0, f/2.8, f4.0, f/5.6, f/8.0, etc…)

5.4 Image Post Processing

The coloured image of the parts is difficult to apply a computer vision algorithm. To increase reliability and success rate of identification, we need to first convert the images to grayscale. Grayscale images outputs the different intensity values of the images and makes it far easier to implement a pattern recognition search algorithm. To further process the image quality, it would be beneficial to add a red, green or blue filter prior to converting the images to grayscale. This filters out certain colours and makes certain objects contrast better with the background. In our application, the green filter was tested to have the best contrast. The grayscale image will then be further refined using the Laplacian of Gaussian process then threshold for a certain intensity value before the Canny Edge Detection algorithm is applied to obtain the silhouette of the object.

5.5 Thresholding

One of the most convenient forms of detecting objects in the region of interest is using intensity thresholding techniques. This thresholds objects of a certain intensity to be considered an object, and the rest are ignored as noise. This is greatly beneficial for the purpose of this object identification scheme as Gold can be deliberately used as a reflective surface to create a high intensity surfaces that will easily differentiate itself from the rest of the parts and be detected as an object of interest. The objects identified will have to satisfy a surface area criteria, where enclosed objects with surface areas smaller or larger than the pre-defined threshold will be omitted and anything that lies within the threshold will be identified as an object.

However, one of the setbacks of the thresholding technique is that it requires a consistent light source to reproduce the same image intensity on the image space. If the light source has a variable intensity output, then this strategy may not be very reliable.

5.6 Pattern Matching

Pattern matching uses a series of thresholding and edge detection algorithms to detect key features of a template. The best edge detection algorithm was found to be the Canny Edge detection which outweighs the rest of the edge detection strategies.

From there, the key features of the template is searched through the region of interest to identify the object. The Labview pattern recognition algorithm uses a fuzzy logic form of statistical correlations between a template image and objects in the image space. If an object in the image space scores higher than a minimum statistical correlation to the template image, it would be recognized as an object and a red bounding box will be drawn on top of it. If the object does not meet the minimum score, it would be neglected. This process is illustrated in Figure 44 and 45.

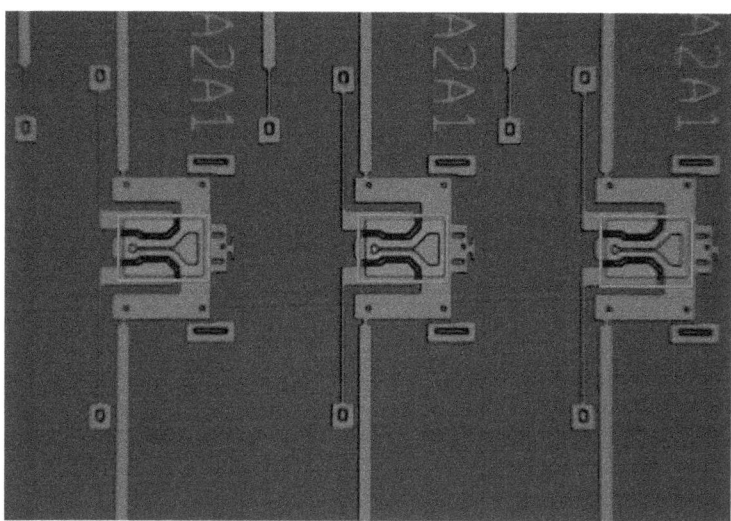

Figure 44: Three Micro-parts identified are in focus and have an average score of 940 with a minimum score requirement of 800

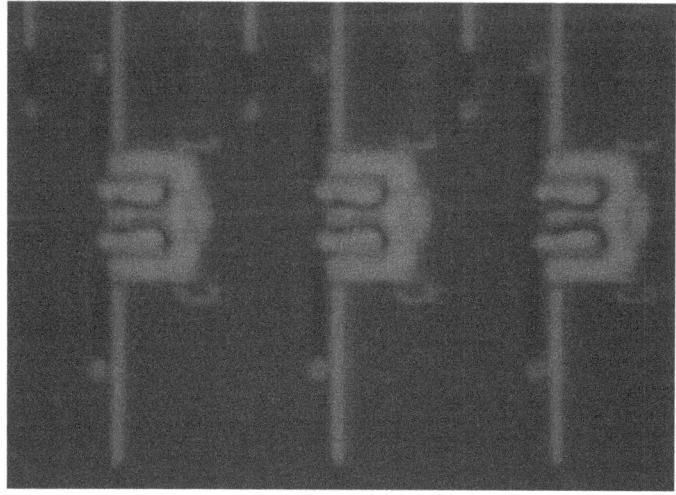

Figure 45: The micro-parts are out of focus and fail to meet the minimum score requirement of 800 and none of them are identified

5.7 Object Identification Success Rate

The object identification process was found to be very high, in most cases 100% as long as the image was in focus. This is because the parts identified are made to very high tolerances and have few or no defects. This makes the computer register a high score for every part on the chip and recognizes it as a part. However, there are certain quirks of the hardware equipment that required to be worked around with. The color intensity of the image plane had the best contrast closest to the center of the image space and gets blurry and distorted towards the edges. Therefore, it is prudent to place the region of interest to close to the center of the image plane of the microscope lens.

5.8 Image Autofocus and Distance Calibration

The microscope has a fixed focal length and a depth of field of 1.5um. Using it as a constraint, we can engineer an autofocus algorithm using contrast values that helps align the micro-gripper to the same plane as the micro-part and facilitate the auto-gripping process.

The autofocus algorithm in the computer vision program is important to bring the lens in focus to the specimen. What is commonly used out there in the industry is to use the highest contrast value in an image as a point of focus.

Most autofocusing algorithms uses the sum of Laplacian of Gaussian (LoG) function as a means to attain the contrast values of an image. It does it by taking the second partial differential values of the image pixel in the x and y direction, which indicates the rate of change of intensity values between adjacent pixel values. An image that is in focus has a sharp contrast, hence a high rate of change of intensity values between adjacent pixels, as opposed to an image that is not in focus which explains the blurred edges.

$$L(x,y) = \frac{\partial^2 I}{\partial x^2} + \frac{\partial^2 I}{\partial y^2}$$

$$\sum L(x,y) = Contrast$$

Figure 46 below summarizes the autofocus algorithm that most conventional autofocusing modules utilize.

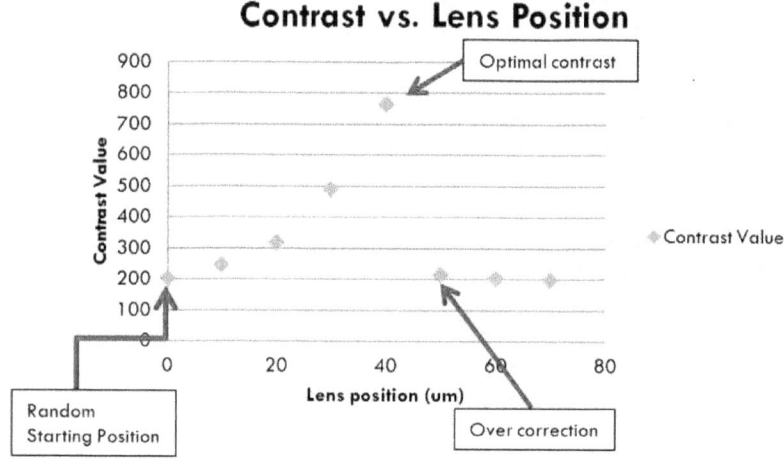

Figure 46: Conventional autofocusing module uses contrast values to gauge the correct lens position

However, as a form of novelty and also simplicity in the programming of the autofocus algorithm, the highest part score from the pattern matching algorithm in Labview was used as a measure for an image in focus. This is illustrated in Figure 47 and 48.

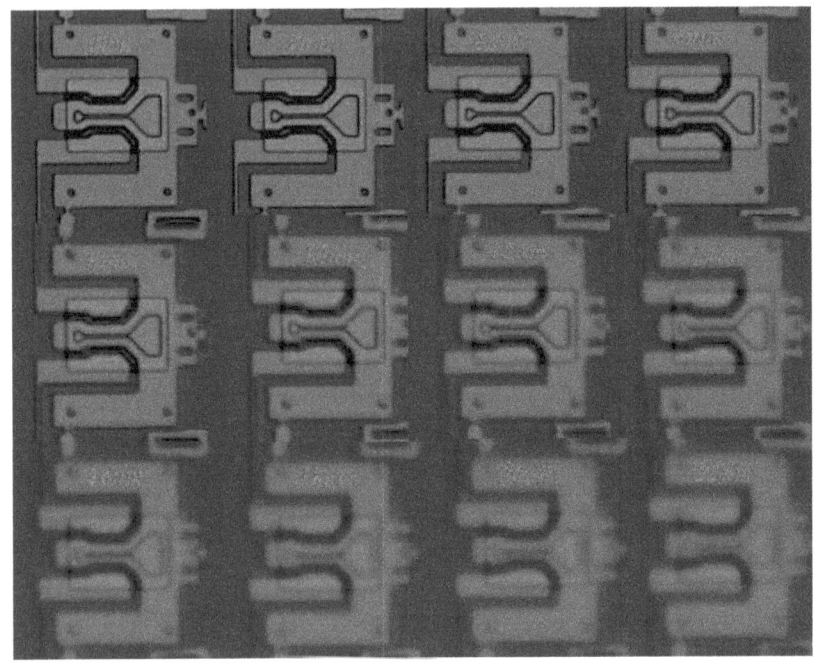

Figure 47: The contrast of the micro-part as the lens moves further away from the focal length

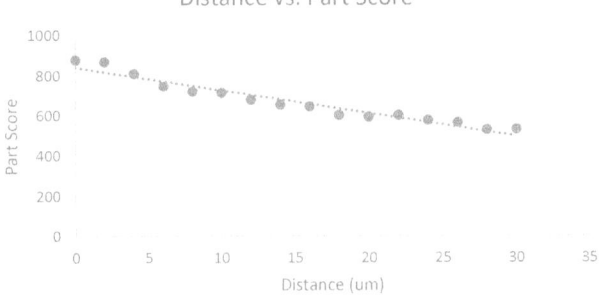

Figure 48: The distance offset away from the focal length to part score follows a strictly linear relationship

It turns out that the part score from the pattern matching algorithm correlates to the distance offset of the lens to the specimen linearly. From this relationship, the use of part scores can be used to estimate the distance offset of the microscope to the specimen and an autofocusing algorithm can be derived from this relationship. The use of the autofocusing algorithm for the use of automated gripping will be discussed in detail in Chapter 7.

Chapter 6

Visual Markers

6.1 Reference Points

In order to create a successful automation process, the system requires visual markers around the chip to serve as reference points for the computer to acquire spatial information of the gripper's position in relation to the chip. This process is analogous to having an origin point in any Cartesian coordinate system problem.

6.2 Design of Visual Markers

The design of the visual markers utilizes a circle design with 3 different sizes of circles. The circle is used as a basic design for object identification because of its incorruptible symmetry regardless of spatial orientation (A circle retains its symmetry regardless of rotation). This prevents image rasterization that is associated with linear objects which would distort the image and reduce the success rate for pattern recognition. Although one may argue that you could run through a series of transforms such as the Hough transform to attain linear information about an object. However, doing more image transformations on an image takes up more computational time, which is crucial for live feedback, and does not necessarily guarantee success. By using circles as a reference geometry, we mitigate the need to use transforms, simplifies the process and guarantees a high success rate. Figures 49 and 50 below illustrate the pixelated edges of a circular object versus a square silhouette.

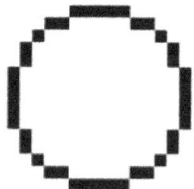

Figure 49: The pixel edge detection of a circle remains circular regardless of rotational offset

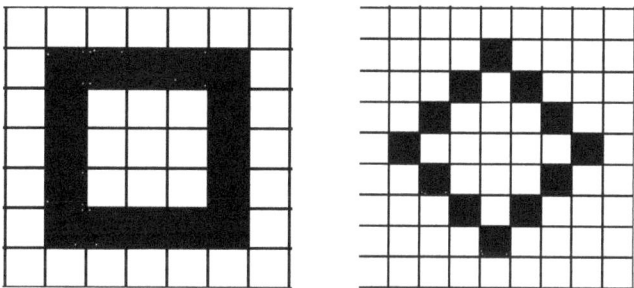

Figure 50: A rectilinear figure such as a square, would be distorted and rasterized as it undergoes rotational offset

After establishing the need to use circles instead of rectilinear objects as a reference geometries, a proposed method of reference geometries for the PolyMUMPS chip is a unique 3 circle system. The 3 circles are made of metal (Gold) which is embedded on a Poly2 and Anchor2 pad as shown in Figure 51 below.

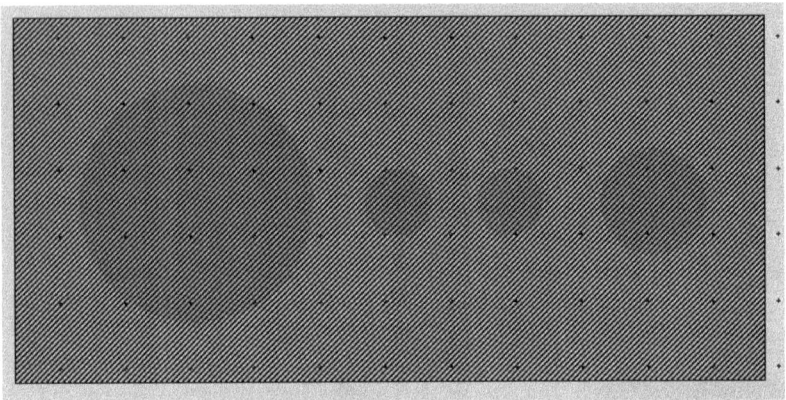

Figure 51: Reference geometry designed on L-edit

The large circle on the left is the anchor point on the reference geometry. The mid-size circle on the extreme right is the extension of the reference geometry that will aid in telling the computer how much the chip is misaligned by. The two small circles in the middle denotes which reference geometry this is on the chip

The large and medium sized circles would determine the chip's orientation angle and determine whether the chip has to be rotated further to attain ideal parallelism. This is done by figuring out the vertical and horizontal displacement of the large and medium circles with respect to each other, and by using trigonometric relations, the corresponding angle between them can be calculated. The smaller circles serves as a means to identify the specific visual marker on the chip. The computer counts the number of small circles there are on the visual marker and the corresponding number of circles will be associated to a specific reference point on the chip, as seen in Figure 52 below.

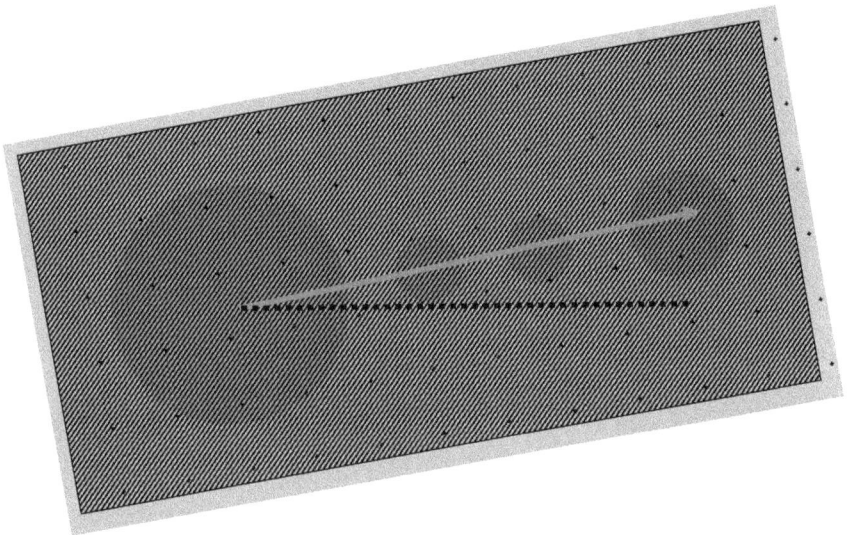

Figure 52: Chip rotational offset

$$Degrees\ Offset = tan^{-1}(\frac{y}{x})$$

The idea of using circles was inspired by how blind people read braille. It is a simple and robust process that has few complications. The way the visual markers are manufactured is by placing a gold layer on top of Anchor1, which is in turn placed on top of the Poly1. In this configuration, the circular marker forms the greatest intensity and contrast values possible in the PolyMUMPS process.

The image identification process can be done in two generic ways. In particular, pattern matching or thresholding. In the pattern matching algorithm, Figure 53, the computer searches for the circular pattern, using a correlation factor, in the image space and anything that scores above a minimum score is recognized as a feature, similar process to recognizing parts on the chip.

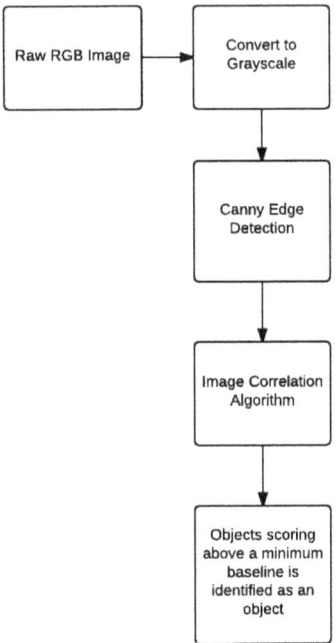

Figure 53: Pattern matching algorithm

In the thresholding process, the computer thresholds intensity values above a certain value and calculates the total enclosed area that feature has and matches it to the area criteria that the user has placed for that particular feature. Figure 54 below shows the flow chart for the vision identification algorithm.

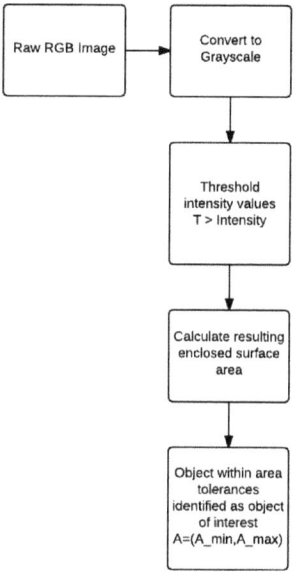

Figure 54: Part recognition using the thresholding algorithm

It was found through experiment and trial and error that the pattern matching method was more reliable and superior to the thresholding technique of recognizing individual circles. This was because the thresholding method was highly susceptible to noise in the image and would falsely recognize other parts in the image space as a circle. Hence, the pattern matching algorithm was used in the routine of recognizing the reference geometries.

6.3 Visual Marker Identification Success Rate

The visual marker identification success rate is very high, no exact statistical number was recorded, but it worked in every time it was utilized properly. In cases which it fails, it was due to a dirt laying on top of the visual marker, causing it to be incorrectly identified. Or in some cases, if the visual marker was damaged the computer will not identify the marker. It is also important to note that to make the identification success rate high, correct saturation, exposure and filter values have to be adjusted for the pattern matching algorithm to work at a high precision level.

6.4 Visual Marker Implementation

The visual markers are placed on all corners of the chip and in between the spaces inside the chip. Ideally, the chip only requires a minimum of 3 visual markers to triangulate the exact position of the end-effector in space. However, for redundancy purposes more markers were placed on the chip substrate.

The visual marker's use in the assembly process goes in a procedural order:

1. The user manually identifies a visual marker and initiates the visual calibration process
2. The computer calculates the angle at which the chip is offset and rotates the chip to orientate it correctly.
3. It does an auto search to the nearest visual marker and identifies it. This will generate a spatial value of the camera with respect to the chip.
4. The computer will than search the corners of the chip to identify the visual markers to make a spatial adjustment across the chip.
5. The camera will move back to the referenced origin of the chip and all the relative X, Y and Z coordinates are zeroed.
6. Automation is ready to proceed.

Chapter 7

Computer Vision and Automation

In order to achieve automation, computer vision is used to feed the computer live coordinate values of where the micro-parts and micro-gripper are in relation to each other. In addition to that, the computer vision acquisition variables have to be fixed and calibrated for reproducibility of results.

A simple flow chart providing an overview of the gripping and joining process developed in this work, is illustrated in Figure 55.

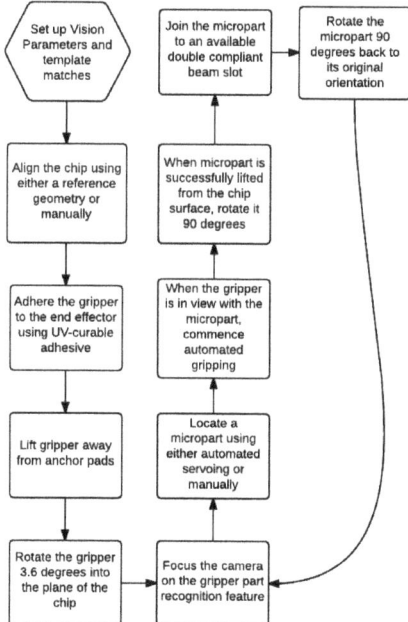

Figure 55: Automation Sequence flow chart

7.1 Pre-requisites to Automation

To create a consistent image capture, and to increase computer vision reliability, we need to fix the variables in the camera settings. It has been found that the following settings produce favourable results: exposure time of 100 ms, 0 gain, 130% saturation, Gamma of 2.2, white balance of (Red = 1, Green = 1.2, Blue = 1.9). The key is not in the numbers chosen for image acquisition, but choosing a fixed and consistent benchmark for acquisition to work in order to create a high reliability for reproducing results. Figure 56 below shows the Pixelink camera settings fixed for the process.

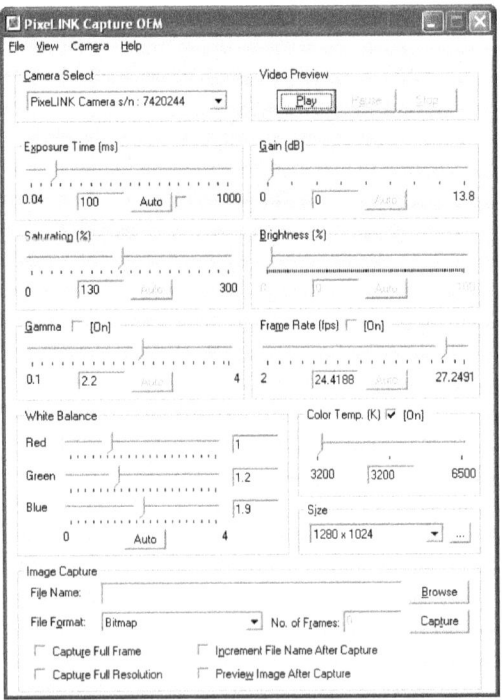

Figure 56: Pixelink Variable Settings

7.1.1 Checklist of Conditions

The following algorithm is used for successful automation:

1) Ensure alignment of parts, using any reference point geometries on the chip (Circle geometries) to correct for misalignment errors, as seen in Figure 57. The computer vision module senses the two geometrical features in the image space, it calculates the angle offset between them, and is corrected using a precise motion control command, as seen in Figure 58.

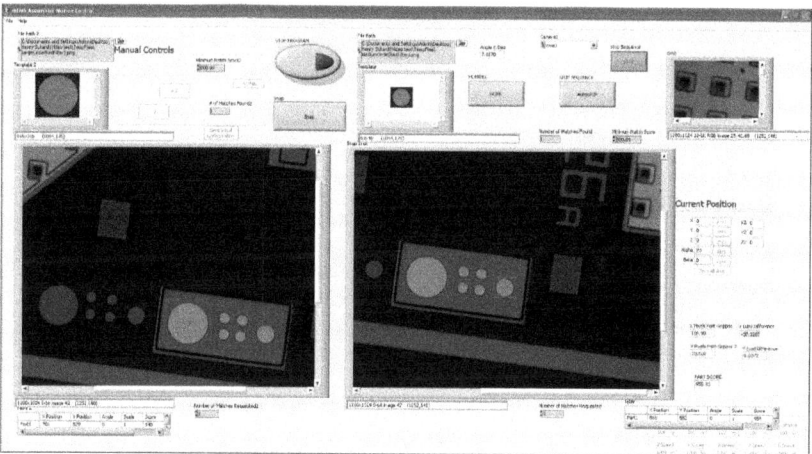

Figure 57: Unaligned geometry detected by computer vision

61

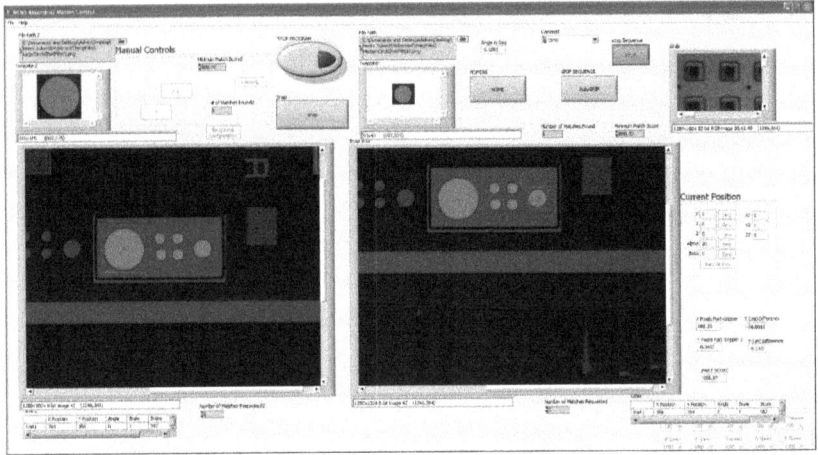

Figure 58: Aligned part geometry using motion control correction

2) The user searches for the nearest available gripper in order to attach it to the needle (end-effector) using a UV-curable adhesive

3) The micro-gripper is lifted away from its anchor

4) The micro-gripper is rotated 3.6 degrees into the plane of the chip

5) The camera is focused on the micro-gripper, specifically on the gripper tips, which is needed for part recognition

6) A micro-part must be located either manually (by user) or using visual servoing to identify a micro-part of interest

7) The micro-part and the micro-gripper plane must be in the same relative image space and image depth, in order for the auto-gripping sequence to work

8) Verify that the computer senses both parts (micro-gripper and micro-part). Initiate auto-grip.

9) If the auto-gripping process is successfully completed, the micro-part is successfully gripped and lifted from the tethers that had anchored it to the surface of the chip

10) The micro-part is rotated 90 degrees

11) The micro-part is joined with a double compliant beam slot.

Visual diagrams of this process are shown in Chapter 2.

7.2 Automation Programming Architecture

The Labview programming architecture consists of two main "while loops" running simultaneously. This gives the motion control a live feedback of what it is doing and whether to continue the next subroutine of code or to keep re-iterating a previous program until a condition is met, as seen in Figure 59.

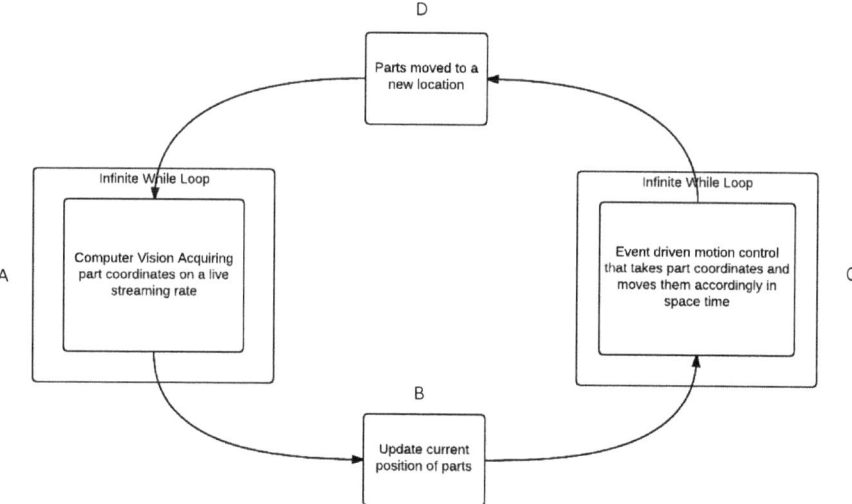

Figure 59: Automation Programming Architecture

The specifics of how the computer vision and motion control works would be reviewed in detail in later sections of this chapter.

7.2.1 Computer Vision Program Module

The way the computer vision program works is by acquiring live images from the Pixelink camera and processing it in a case structure with a sampling rate of 100 ms per image or 10 times a second. In the case structure, it does a Green filter to attain the grayscale image and then takes an image template and searches for the template in the image space. If an object that exceeds the minimum part score is found, it would be identified by a bounding red rectangle in the image space.

A simplified flow chart is shown in Figure 60 below on the computer vision process.

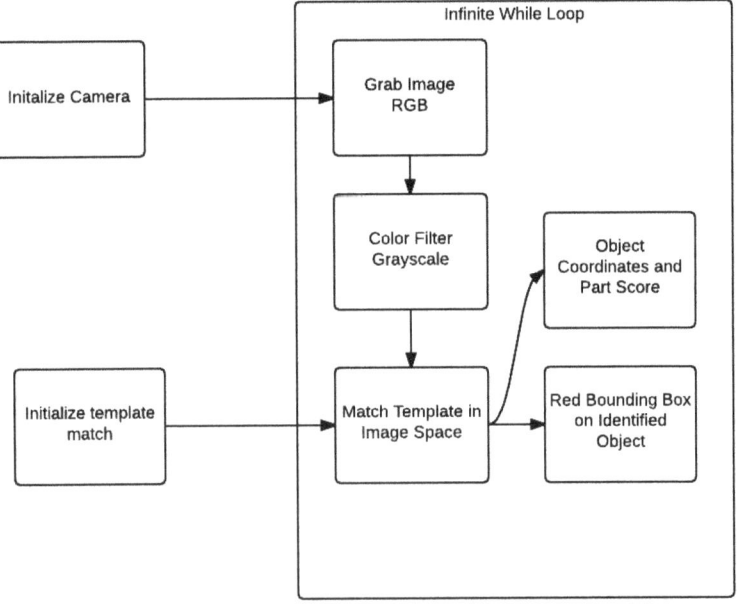

Figure 60: Computer Vision Architecture

Figure 61 below shows the actual computer vision program done in Labview.

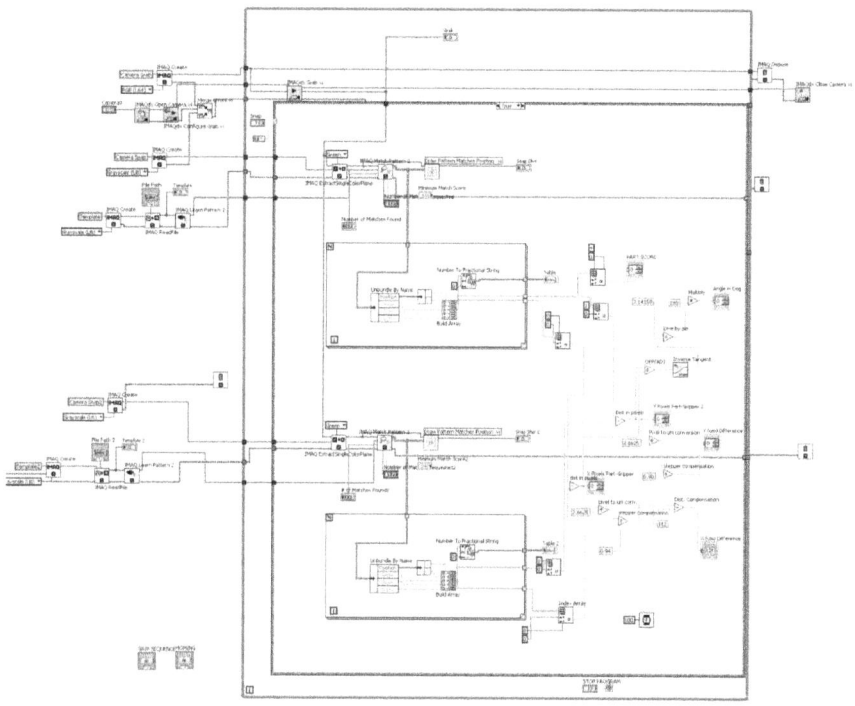

Figure 61: Labview Computer Vision Module

7.2.2 Autofocus and Pattern Matching Score
The diagram in Figure 62 below shows the flow chart of the autofocusing process.

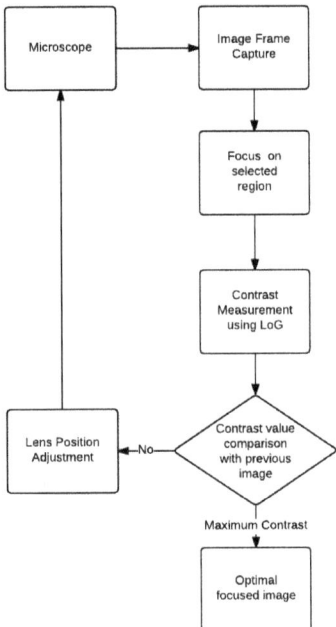

Figure 62: Conventional Autofocus Algorithm

In this computer vision program, another means of approximating the focus of the camera lens with respect to the chip parts was done by comparing the score of the part with respect to the template. It was deduced from empirical testing that with every 1 um offset from the focal length of the microscope, the part score would lose approximately 20 points. This trend however, was only linear for the first 10 um, after which the scoring to distance of the parts become exponentially inaccurate. A part that is in focus would score between 940-960 points out of a total of 1000, whereas a part that is 10 um out of focus would attain between 740-760 points out of a score of 1000.

Figure 63 below summarizes a flow chart for this process.

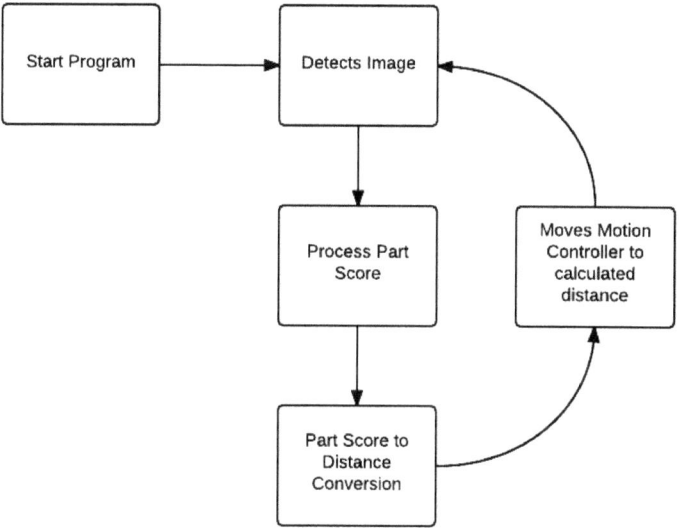

Figure 63: Part score for distance calibration

The Labview pattern matching algorithm uses a fuzzy logic method of comparing an object to the template of interest. This correlates the object similarity to the template and scores the identified object with respect to the template out of a score of 1000. Using this pattern matching algorithm, we can use it to identify objects of interest and gauge the focal distance to the chip plane to a certain extent by judging the part score. A detailed description of how this works was discussed in Chapter 5.

7.2.3 Motion Control Program

The way the motion control program works is by taking the live part coordinates calculated by the computer vision portion of the program and run it through its own sequence of motions. The most robust portion of this program is that the computer vision program constantly feeds updated coordinate values of the parts in the image space to the motion controller, so as to give accurate judgement on whether to proceed to the next set of movements or to re-iterate and adjust itself first, as illustrated in Figure 64.

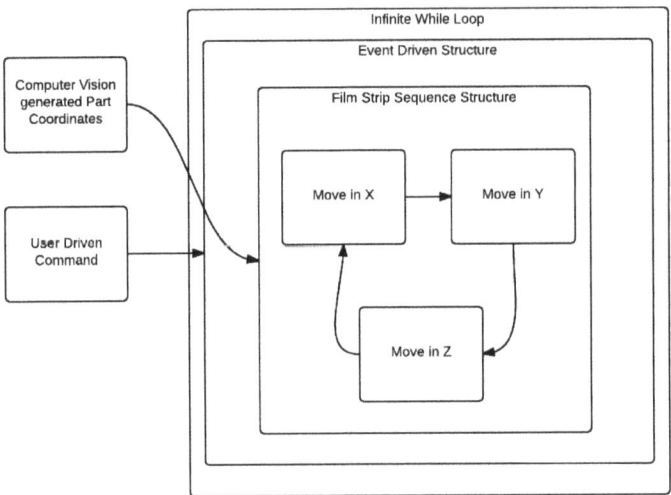

Figure 64: Motion Control Program

The film strip sequence structure does a sequential motion in the X, Y and Z axis. After each motion, the computer vision updates the part coordinate values in the image space before proceeding to the next sequence in the film strip.

The Figures 65 and 66 below shows a sample of several pieces of motion codes used to join the gripper to the micro-parts.

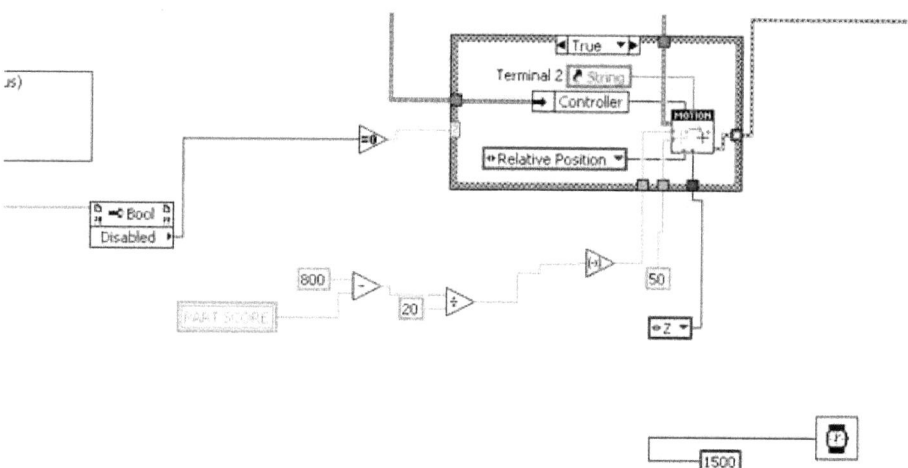

Figure 65: Labview Single Axis Motion Execution

The diagram below shows the nested "while loop" process in the film strip sequence that iterates a sequence of motion continuously until a computer vision calculated coordinate system is met.

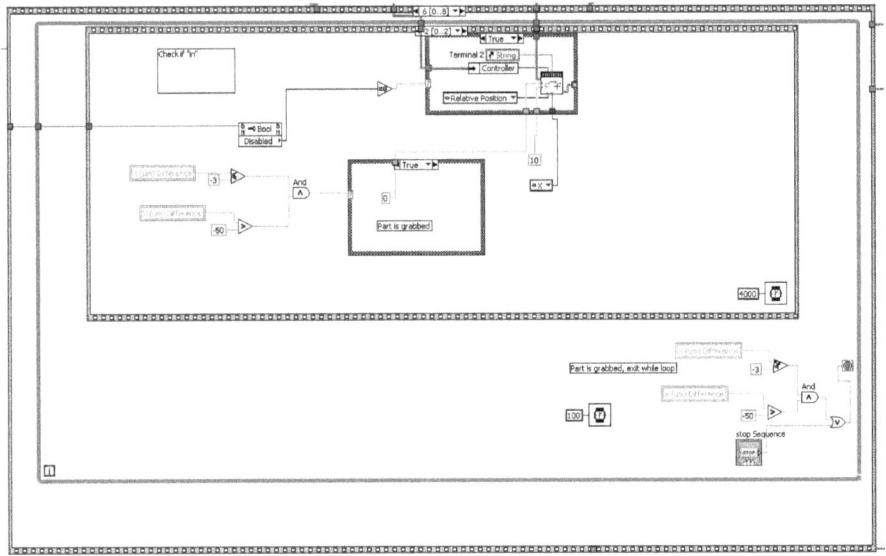

Figure 66: Nested while loop, conditional motion control

Chapter 8

Results

To test the capability of the computer vision program for automated assembly, two parts were placed in the search variable, which were the micro-part and the micro-gripper. This makes the computer vision module search for these two parts in the image space as shown in Figure 67.

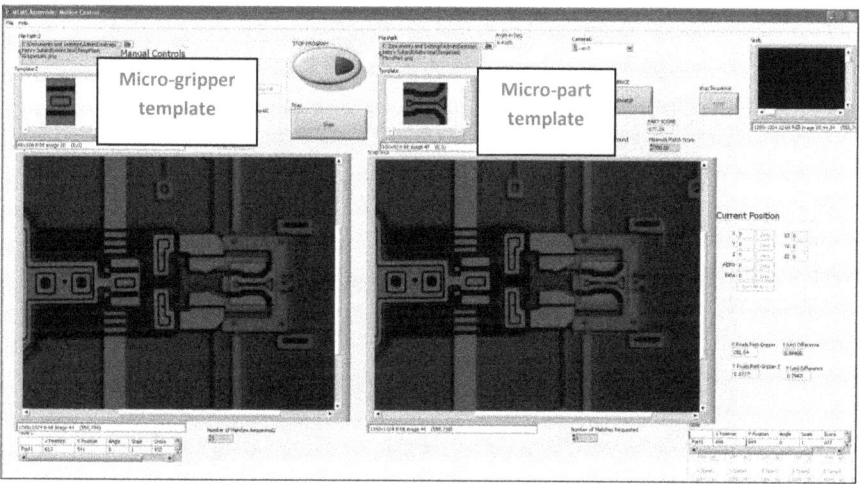

Figure 67: Two Separate templates are being searched in the image space and identified using a bounding red box

To ensure reproducibility of the experiment, the image acquisition parameters were fixed. The variables fixed include the saturation values, exposure time, gain, white balance, color temperature, and the rate of frames acquired per second.

To test the robustness of the auto-gripping algorithm, the micro-gripper was placed in random locations in the image space with varying X, Y and Z offsets. This is to test whether the code would work under different situations and at which point the program would fail to properly execute the program. The coordinate offsets were recorded for each trial and ran for a total of 10

trials, as illustrated in Figure 68.

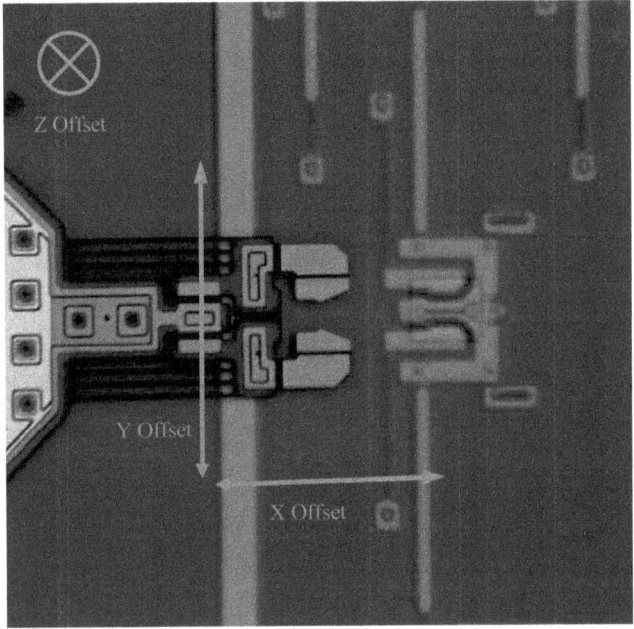

Figure 68: The micro-gripper is placed in random X, Y and Z offsets away from the micro-part for each trial before the automated gripping is initialized

The experiment was performed 10 times using different values for coordinate offsets. The

experiment recorded results for success or failure, number of iterations required to complete a

grasp, the time taken for the grasp process, and a conclusion on the process. Tabke 1 below

summarizes the experimental results.

Table 1: Experiment results of 10 trials of the automated gripping

Trial#	X-Offset	Y-Offset	Z-Offset	Outcome	Iterations	Time (s)
1	25	20	10	Successful	1	30
2	33	9	8	Successful	2	36
3	21	13	5	Successful	2	36
4	24	-18	12	Successful	1	31
5	22	21	4	Successful	2	38
6	21	9	7	Successful	4	57
7	24	-18	11	Successful	1	30
8	34	5	5	Successful	2	38
9	57	14	7	Successful	9	86
10	29	-18	13	Successful	2	37

The average time for automated gripping is 30 seconds for a single iteration direct grab, and 37 seconds for two iterations. For every additional iteration required, 7 seconds are added. A majority of the trials required two iterations, while some trails required 4 iterations, and an anomaly of 9 iterations was reached in one instance.

In comparison, the average time for a skilled person to complete tele-operated gripping is 60 seconds. This means that this automated gripping process provides time savings in being able to grab and hold micro-parts, by being faster than a human user.

The current success rate for the automated gripping, using the same conditions for every trial is 10 success out of 10 trials. Although the sample data is not sufficient to show that it offers a 100% success rate, it does show the process is very robust and has a high success rate.

Chapter 9

Discussion, Recommendations and Future Work

9.1 Discussion

The work that has been done thus far will lay foundations to future micro-assembly work. As of now, the micro-gripping process has been automated, and resulted in a high success rate with 10 successes out of 10 trials.

The chip fabrication process went very successful and there was few or no errors in the fabrication process and the chip parts were made to very high tolerances. However, a small error involving a design flaw was incorporated in the chip which should be avoided in the future. There are Poly1 layers placed across the circumference of the chip which was meant to be used as a visual boundary for the fabrication process, however this Poly1 layer impeded the gripping process as it was too close to the micro-parts close to the edge of the chip. Future designs should eradicate the use of this Poly1 layer around the chip.

The Labview software developed works well as long as the software does what it was meant to do. The Labview program breaks down if the micro-gripper fails to contact the micro-part on the first attempt or slips in an iteration. This causes the computer vision module to think that the micro-gripper is inside the micro-part and the joining process should proceed as usual. Although this scenario is highly unlikely, because the system has been calibrated to a high accuracy, but was proven to be a problem when the system was deliberately not well calibrated. In future work, perhaps a more robust subroutine should be implemented between processes to check for gripping slippage. It is also possible to implement a computer troubleshooting heuristic process of assessing the problem and re-calibrating its motion values on the next attempt.

9.2 Recommendations and Future Work

Future work would involve automating the joining process. This would require the assembly of micro-parts into the slots. This is not a trivial feat as the features become blurred (due to the low depth of field) and the part recognition process becomes more intricately involved. This would require an intensive autofocusing algorithm to attain good estimation of vertical distance, and feedback control. Such a task is difficult even for a human user to perform joining. A possible use of microscopy image de-convolution at an angle to the surface of the chip could be used to estimate the joining procedure and distance estimation more accurately than the current planar depth of field. The way image de-convolution works is by taking numerous planar images obtained by the microscope, and stacking them on top of each other for images at different depths. This can produce a clear, in-focus image of the specimen without the use of an SEM (Scanning Electron Microscope).

The computer vision algorithm right now does automation in segments, meaning the homing procedure, angle correction, part finding and automated gripping are all separated systems. This was done to test the robustness of the individual codes created before they will be implemented as a complete process. What future work entails is taking all the codes developed in this report to be stitched together and create a fully automated assembly process with a single click of a button. This means that at the start of the program, the user sets the variables of the template images such as the micro-part, micro-gripper and circle geometries. Once the end-effector is adhered to a micro-gripper, the user can just initiate the program sequence and the computer will take full responsibility of searching for an available micro-part, re-orientate the micro-gripper in space and grip the micro-part out of its slot.

Chapter 10

Conclusion

The intention of the author was to develop an autonomous MEMS vision and manipulator system with the following objectives: (i) develop a new generation MEMS chip that incorporates new micro-parts and features meant for automated gripping and assembly, (ii) develop a new five degree-of-freedom robot integrated with a three degree-of-freedom optical system capable of micromanipulation, autofocusing and visual servoing, (iii) develop a computer vision program that will utilize object recognition and motion control synchronously to enable a fully automated system

The MEMS chip was designed with new features and parts, and was successfully fabricated and tested in this work. The new MEMS chip parts were found to be superior in assembly robustness and reliability to the predecessors, leading to much higher overall success rate in micro-assembly. The chip has also incorporated visual markers across the chip, to enable auto-alignment and visual servoing. The parts were fabricated in a fashion to allow for a streamlined assembly process, and to allow for improved turnaround time for successfully assembled micro-parts.

The new 5 DOF robot for MEMS micromanipulation was not fully assembled and tested for use. However, numerous parts were redesigned and built to be assembled in the near future. Due to certain bottlenecks of the development process, the older existing robotic was used to test out the computer vision program and the MEMS chip.

The computer vision program was successfully developed in Labview and works with a high success rate in automated gripping. The motion controller and computer vision program worked very well together to effectively recognize objects in the image space, to adjust the focus

by changing the distance in the Z axis, and to align parts in the image space in the X and Y axis

for automated gripping.

The successful integration of a five degree-of-freedom robotic system, the new MEMS

chip design, and the developed computer vision program, will lay the foundations for future

work. The ultimate goal is a fully autonomous system that will use visual servoing to identify

parts on the chip, auto-grip and auto join the parts to the allocated slots, where the human user is

only required to input high level commands to initiate the program.

Bibliography

[1] I. B. Bahadur, J. Mills and Y. Sun, "Design of a MEMS-Based Resonant Force Sensor for Compliant, Passive Micro-gripping," *IEEE International Conference on Mechatronics & Automation,* 2005.

[2] M. Basha, N. Dechev and Naeini, Safieddin, Chaudhuri, Sujeet, "Improved Design of Large 3D Micromirrors for Micro-assembly onto an Optical MEMS Cross-COnnect," *SPIE 6717, Optomechatronic Micro/Nano Devices and Components,* 2007.

[3] K. L. Chen, "Novel MEMS Grippers For Pick-Place of MicroAnd Nano Objects," *University of Toronto,* 2009.

[4] N. Dechev, W. L. Cleghorn and J. K. Mills, "Development of a 6 Degree of Freedom Robotic Micromanipulator for Use in 3D MEMS Micro-assembly," *IEEE International Conference on Robotics and Automation Orlando, Florida,* 2006.

[5] N. Dechev, W. L. Cleghorn and J. K. Mills, "Design of Grasping Interface for Micro-grippers and Micro-Parts Used in the Micro-assembly of MEMS," *IEEE International Conference on Information Acquisition,* 2005.

[6] N. Dechev, W. L. Cleghorn and J. K. Mills, "Micro-assembly of 3-D Microstructures Using a Compliant, Passive Micro-gripper," *Journal of Microelectromechanical Systems,* vol. Vol. 13, 2004.

[7] N. Dechev, W. L. Cleghorn and J. K. Mills, "Tether and Joint Design for Microcomponents used in Micro-assembly of 3D Microstructures," *MEMS/MOEMS Components and their Applications,* 2004.

[8] N. Dechev, W. L. Cleghorn and J. K. Mills, "Micro-assembly of 3-D MEMS Structures Utilizing a MEMS Micro-gripper with a Robotic Manipulator," *IEEE International Conference on Robotics & Automation Taipei, Taiwan,* 2003.

[9] N. Dechev, W. L. Cleghorn and J. K. Mills, "Construction of 3D MEMS Microstructures Using Robotic Micro-assembly," *Department of Mechanical and Industrial Engineering, University of Toronto,* .

[10] B. Mohamed, N. Dechev, S. S Naeini and SK Chaudhuri, "Digital Optical 1×N MEMS Switch Utilizing Microassembled Rotating Micromirror," *Optical MEMS and their Applications Conference,* 2006.

[11] G. Sharma, S. Sood, G. S. Gaba and N. Gupta, "Image Recognition System using Geometric Matching and Contour Detection," *International Journal of Computer Applications,* vol. Vol 51, 2012.

[12] R. Szeliski, Ed., *Computer Vision: Algorithms and Applications*. 2010.

[13] N. Dechev, "MECH 466 - Microelectromechanical Systems. Lecture Notes and Lab Manual," *University of Victoria*, 2014.

[14] M. A. Greminger, A. S. Sezen and B. J. Nelson, "A Four Degree of Freedom MEMS Microgripper with Novel Bi-Directional Thermal Actuators," *IEEE/RSJ International Conference on Intelligent Robots and Systems*, 2005.

[15] G. A. Singh, D. Horsely, M. Cohn, A. Pisano, and R. Howe, "Batchctransfer of microstructures using flip-chip solder bonding," *J. Microelectromech. Syst.*, vol. 8, pp. 27–33, Mar. 1999.

[16] M. M. Maharbiz, R. T. Howe, and K. S. J. Pister, "Batch transfer assembly of micro-components onto surface and SOI MEMS," in *Proc. Transducers '99 Conference*, Sendai, Japan, June 7–10, 1999.

[17] T. Ebefors, J. Ulfstedt-Mattsson, E. Kälvesten, and G. Stemme, "3D micromachined devices based on polyimide joint technology," in *SPIE Symposium on Microelectronics and MEMS*, vol. SPIE Vol. 3892, Gold Coast, Queensland, Australia, October 1999, pp. 118–132.

[18] K. F. Harsh, V. M. Bright, and Y. C. Lee, "Solder self-assembly for threedimensional microelectromechanical systems," *Sens. Actuators, Phys. A*, vol. 77, pp. 237–244, 1999.

[19] J. Zou, J. Chen, C. Liu, and J. E. Schutt-Ainé, "Plastic deformation magnetic assembly (PDMA) of out-of-plane microstructures: technology and application," *J. Microelectromech. Syst.*, vol. 10, pp. 302–309, June 2001.

[20] K. F. Bohringer, K. Goldberg, M. Colm, R. Howe, and A. Pisano, "Parallel microassembly with electrostatic force fields," in *Proc. International Conference on Robotics and Automation (ICRA98)*, Leuven, Belgium, May 1998.

[21] S. R. Burgett, K. S. J. Pister, and R. S. Fearing, "Three dimensional structures made with microfabricated hinges," in *Proc. MicroMechanical Systems, ASME Winter Annual Meeting*, vol. 40, Anaheim, CA, 1992, pp. 1–11.

[22] A. Friedberger and R. S. Muller, "Improved surface-micromachined hinges for fold-out structures," *J. Microelectromech. Syst.*, vol. 7, pp. 315–319, Sept. 1998.

[23] E. Hui, R. T. Howe, and S. M. Rodgers, "Single-step assembly of complex 3-D microstructures," in *Proc. IEEE Thirteenth Annual International Conference on Micro Electro Mechanical Systems*, Miyazaki, Japan, Jan 2000, pp. 602–607.

[24] E. Shimada, J. A. Thompson, J. Yan, R.Wood, and R. S. Fearing, "Prototyping millirobots using dextrous microassembly and folding," in *Proc. ASME IMECE/DSCD*, Orlando, Florida, Nov. 5–10, 2000, pp. 933–940.

[25] G. Yang, J. A. Gaines, and B. J. Nelson, "A flexible experimental workcell for efficient and reliable wafer-level 3D microassembly," in *Proc. IEEE International Conference on Robotics and Automation (ICRA 2001)*, Seoul, South Korea, 2001, pp. 133–138.

[26] Zyvex Microgrippers, Micromachined Silicon Structures, Top Down Group.*http://www.zyvex.com/Products/Grippers.html* [Online]

[27] C. G.Keller and M. Ferrari, "Milli-scale polysilicon structures," in *Proc. Solid-State Sensor and ActuatorWorkshop*, Hilton Head, SC, June 1994, pp. 132–137.

[28] Photo Gallery, MEMS Precision Instruments, C. G. Keller.*http://www.memspi.com/gallery.html* [Online] DECHEV *et al.*: MICROASSEMBLY OF 3-D MICROSTRUCTURES USING A COMPLIANT, PASSIVE MICROGRIPPER 189

[29] J.T. Feddema and R.W. Simon, "Visual servoing and CAD-driven microassembly," *Special Issue on Visual Servoing, IEEE Robotics and Automation Magazine*, vol. 5, no. 4, pp. 17–24, Dec. 1998.

[30] C. J. Kim, A. P. Pisano, and R. S. Muller, "Silicon-processed overhanging microgripper," *J. Microelectromech. Syst.*, vol. 1, pp. 31–36, Mar. 1992.

[31] N. Dechev, W. L. Cleghorn, and J. K. Mills, "Microassembly of 3-D MEMS structures utilizing a MEMS microgripper with a robotic manipulator," in *Proc. IEEE International Conference on Robotics and Automation (ICRA 2003)*, Taipei, Taiwan, Sept 14–19, 2003.

[32] N. Dechev, W. L. Cleghorn, and J. K. Mills, "Micro-assembly of microelectromechanical components into 3-D MEMS," *Canadian J. Elec. Comput. Eng.*, vol. 27, no. 1, pp. 7–15, January 2002.

[33] S. J. Ralis, B. Vikramaditya, and B. J. Nelson, "Micropositioning of a weakly calibrated microassembly system using coarse-to-fine visual servoing strategies," *IEEE Trans. Electron. Packag. Manufact.*, vol. 23, Apr. 2000.

[34] D. Koester, A. Cowen, R. Mahadevan, and B. Hardy, "PolyMUMPs design handbook revision 8.0," in *MEMSCAP, MEMS Business Unit (CRONOS)* Research Triangle Park, NC, 2001.

[35] W. N. Sharpe and K. Jackson, "Tensile testing of MEMS materials," in *Proc. 2000 SEM IX International Congress*, Orlando, FL, June 5–8, 2000.

[36] W. N. Sharpe, B. Yuan, R. Vaidyanathan, and R. L. Edwards, "Measurements of Young's modulus, Poisson's ratio, and tensile strength of polysilicon," in *Proc. Tenth IEEE InternationalWorkshop on Microelectromechanical Systems*, Nagoya, Japan, 1997, pp. 424–429.

[37] A. A. Yasseen, J. Mitchell, T. Streit, D.A. Smith, M. Mehregany, "A rotary electrostatic micromotor 1□8 optical switch", *Proceedings of The Eleventh Annual International Workshop on Micro Electro Mechanical Systems (MEMS 98)*, Jan 25-29, 1998, pp.116 – 120

[38] T. Ebefors, J. Ulfstedt-Mattsson, E. Kälvesten, and G. Stemme, "3D micromachined devices based on Polyimide Joint technology", presented at *SPIE Symposium on Microelectronics and MEMS*, Gold Coast, Queensland, Australia, SPIE vol. 3892, Oct. 1999, pp. 118-132.

[39] J. Zou, J. Chen, C. Liu, and J. E. Schutt-Ainé, "Plastic Deformation Magnetic Assembly (PDMA) of Out-of-Plane Microstructures: Technology and Application", *Journal of Microelectromechanical Systems,* vol. 10, no. 2, June 2001, pp. 302-309.

[40] D. Koester, A. Cowen, R. Mahadevan, M. Stonefield, and B. Hardy, "PolyMUMPs Design Handbook Revision 9.0", *MEMSCAP, MEMS Business Unit (CRONOS),* Research Triangle Park, N.C., USA, 2001.

[41] M. A. Basha, N. Dechev, S. Chaudhuri, and S. Safavi-Naeini, "Digital Optical 1xN MEMS Switch Utilizing Microassembled Rotating Micromirror", *Proc. of IEEE Optical MEMS 2006*, Aug 21-24, Big Sky, Montana, 2006.

[42] M. A. Basha and S. Safavi-Naeini. "Optimization of electrostatic sidedrive micromotor torque using new rotorpole-shaping technique", *Proc. of SPIE Photonics East*, volume 6374, Boston, October 2006.

[43] Y. Zhou, B. J. Nelson, and B. Vikramaditya, "Integrating Optical Force Sensing with Visual Servoing for Microassembly," *Journal of Intelligent and Robotic Systems*, Vol. 28, pp. 259-276, 2000.

[44] J. Sin, W. H. Lee, D. Popa, and H. E. Stephanou, "Assembled Fourier Transform Micro-spectrometer," *SPIE - The International Society for Optical Engineering*, Vol. 6109, pp. 610904, San Jose, CA, US. Jan. 2006.

[45] J. A. Thompson, and R. S. Fearing, "Automating Microassembly with Ortho-tweezers and Force Sensing," *Proceedings of the IEEE International Conference on Intelligent Robots and Systems*, Vol. 3, pp. 1327-1334, Maui, HI, Oct.-Nov. 2001.

[46] A. Ferreira, C. Cassier, and S. Hirai, "Automatic Microassembly System Assisted by Vision Servoing and Virtual Really," *IEEE/ASME Transactions on Mechatronics*, Vol. 9, No. 2, pp. 321-333, Jun. 2004.

[47] W. Zesch, and R. S. Fearing, "Alignment of Microparts Using Force Controlled Pushing," *Proceedings of SPIE – The International Society for Optical Engineering*, Vol. 3519, pp. 148-156, Nov. 1998.

[48] L. Wang, J. K. Mills, and W. L. Cleghorn, "Adhesive Mechanical Fastener Design for Use in Microassembly," *Proceedings of IEEE - Canadian Conference on Electrical and Computer Engineering*, pp. 1093-1096, Niagara Falls, ON,Canada, May 2008

[49] R. Prasad, K. F. Bohringer, and N. C. MacDonald, "Design, Fabrication, and Characterization of SCS Latching Snap Fasteners for Micro Assembly," *Proceedings of ASME International Mechanical Engineering Congress and Exposition*, Vol. 57-2, pp. 917-923, San Francisco, CA, US., Nov. 1995

[50] L. Zhou, J. M. Kahn, and K. S. Pister, "Corner-Cube Retroreflectors Based on Structure-Assisted Assembly for Free-Space Optical Communication," *Journal of Microelectromechanical Systems*, Vol. 12, No. 3, pp. 233-242, Jun. 2003

[51] L. Ren, L. Wang, J. K. Mills, and D. Sun, "Automatic Microassembly by Vision-Based Control," *Proceedings of the IEEE/RSJ International Conference on Intelligent Robots and Systems*, San Diego, CA, US, Oct.-Nov. 2007

[52] R. S. Fearing, "Survey of sticking effects for micro parts handling," *Proceedings of the IEEE/RSJ International Conference on Intelligent Robots and Systems.*, Vol. 2, pp. 212-217, Aug. 1995

[53] National Instrument, *NI Vision Development Module,* http://www.ni.com/vision/vdm.htm

[54] A.A. Yassen, J. Mitchell, T. Streit, D.A. Smith, and M. Mehregany, "A Rotary Electrostatic Micromotor 1_8 Optical Switch," *IEEE Journal of Selected Topics in Quantum Electronics.*, Vol. 5, no. 1, pp. 26-32, Jan.-Feb. 1999 550

APPENDICES

Appendix A – Labview Program

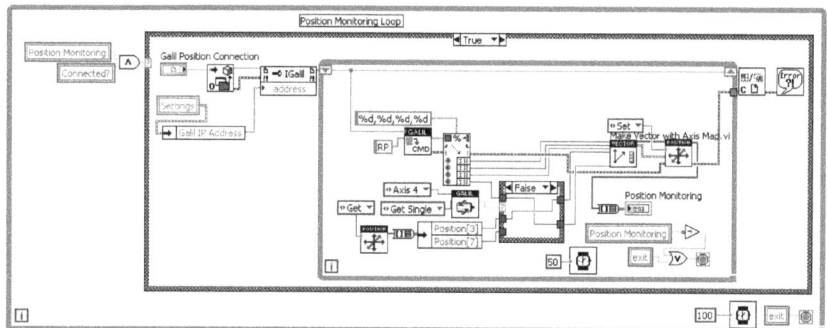

Figure 69: Position Control Loop

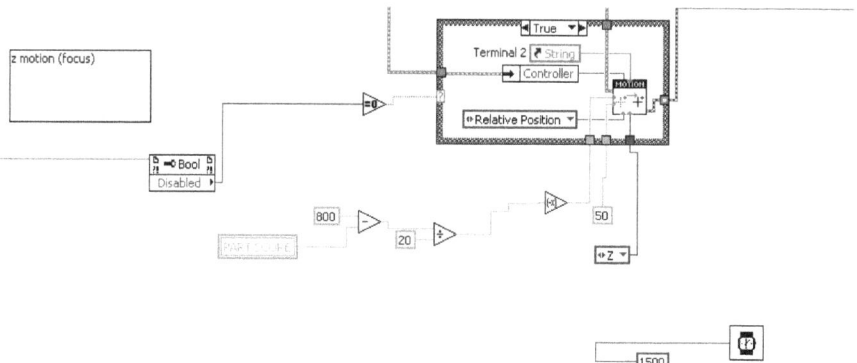

Figure 70: Autofocusing algorithm in the Z-axis

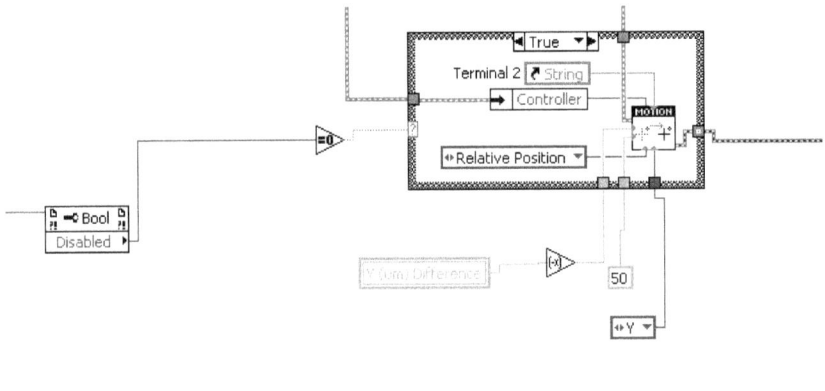

Figure 71: Align in the Y-axis

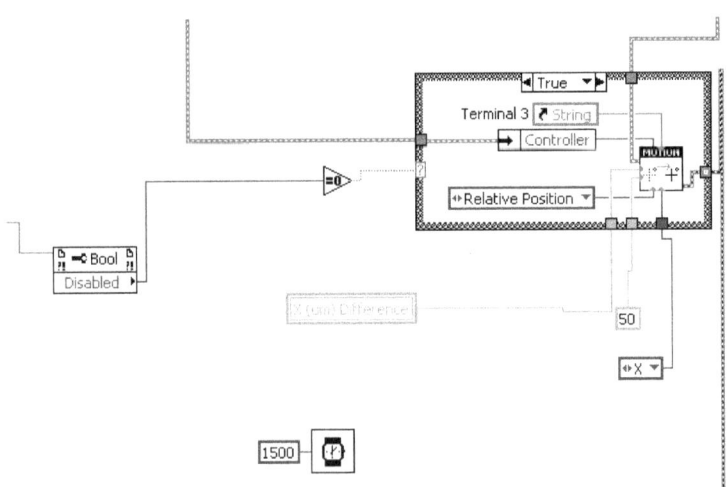

Figure 72: Align in the X-axis

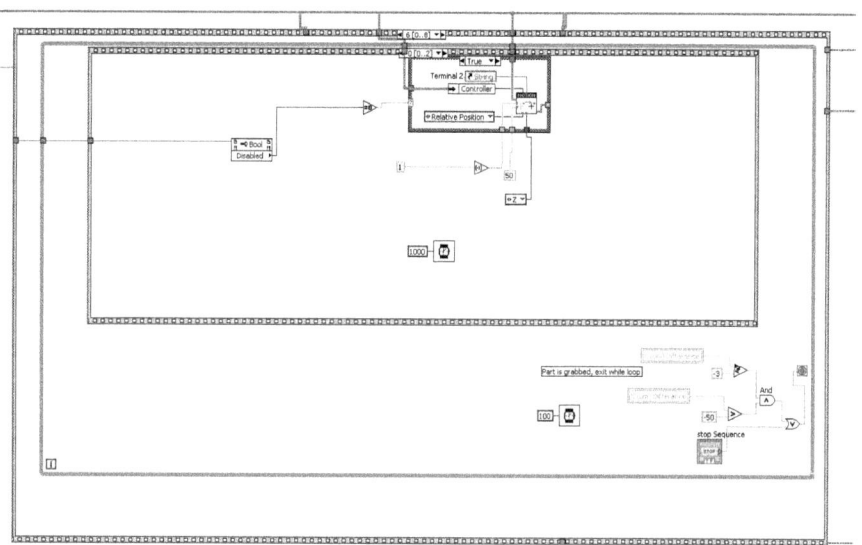

Figure 73: Nested while loop, iterate until the condition is fulfilled

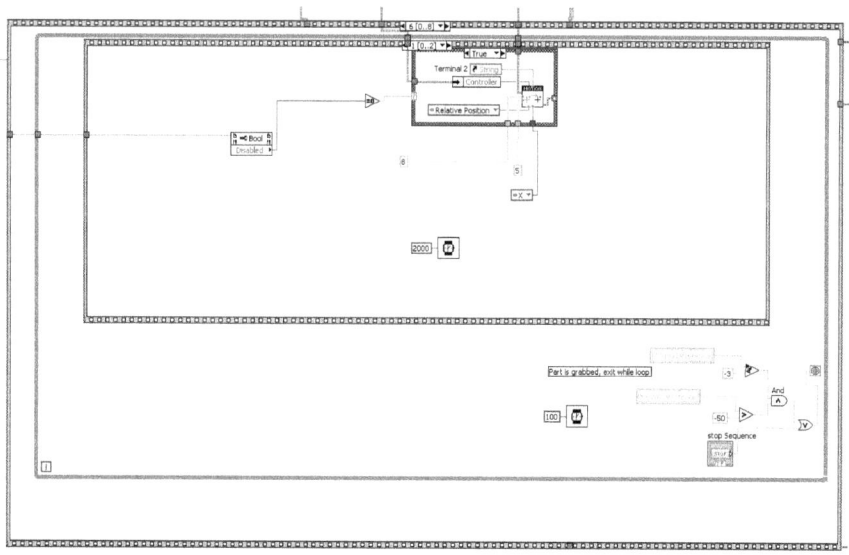

Figure 74: Re-engage in the gripping process again

85

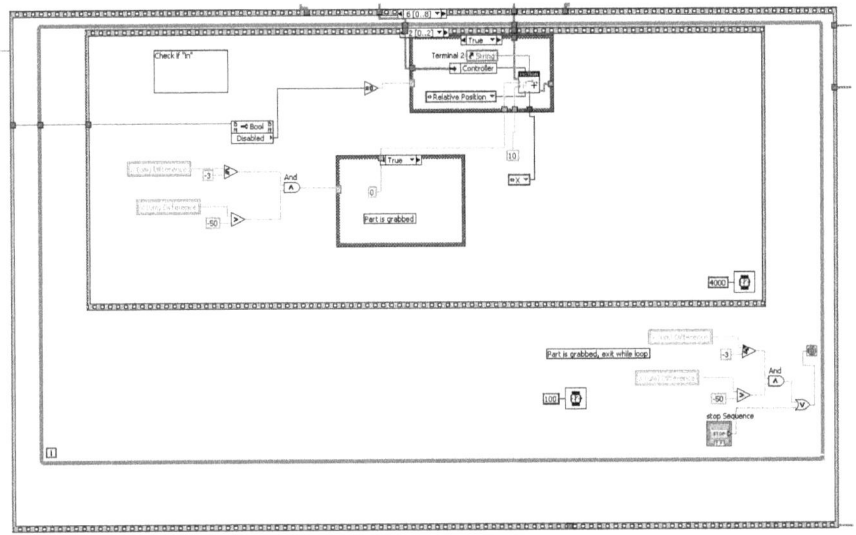

Figure 75: Conditional statement for part being grabbed or re-iterate sequence

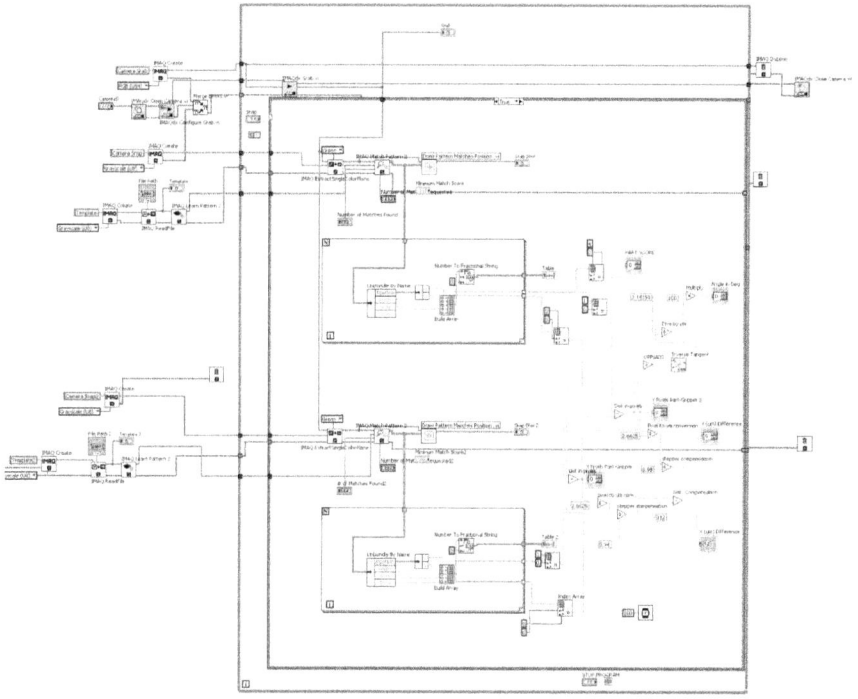

Figure 76: Computer vision position monitoring Loop and pixel to motion control conversion

Lightning Source UK Ltd.
Milton Keynes UK
UKHW011244210119
335934UK00001B/120/P